VOLUME EIGHTY THREE

# ADVANCES IN MARINE BIOLOGY

## Sharks in Mexico: Research and Conservation Part A

# ADVANCES IN MARINE BIOLOGY

**Series Editor**

CHARLES SHEPPARD
*Emeritus Professor, School of Life Sciences*
*University of Warwick, U.K.*
*Charles.Sheppard@Warwick.ac.uk*

VOLUME EIGHTY THREE

# ADVANCES IN MARINE BIOLOGY

## Sharks in Mexico: Research and Conservation Part A

Edited by

**SHAWN E. LARSON**
*Seattle Aquarium,*
*Department of Conservation Programs and Partnerships,*
*Seattle, WA, United States*

**DAYV LOWRY**
*Washington Department of Fish and Wildlife,*
*Fish Management Division, Marine Fish Science Unit,*
*Olympia, WA, United States*

ELSEVIER

ACADEMIC PRESS
An imprint of Elsevier

Academic Press is an imprint of Elsevier
125 London Wall, London, EC2Y 5AS, United Kingdom
The Boulevard, Langford Lane, Kidlington, Oxford OX5 1GB, United Kingdom
525 B Street, Suite 1650, San Diego, CA 92101, United States
50 Hampshire Street, 5th Floor, Cambridge, MA 02139, United States

First edition 2019



ISBN: 978-0-08-102916-9
ISSN: 0065-2881

For information on all Academic Press publications
visit our website at https://www.elsevier.com/books-and-journals

*Publisher:* Zoe Kruze
*Acquisition Editor:* Jason Mitchell
*Editorial Project Manager:* Leticia M. Lima
*Production Project Manager:* Denny Mansingh
*Cover Designer:* Miles Hitchen

Typeset by SPi Global, India

# Contributors to Volume 83


**José Leonardo Castillo-Geniz**
Instituto Nacional de Pesca y Acuacultura, National Fisheries and Aquaculture Institute, Centro Regional de Investigación Pesquera Ensenada, La Paz, Mexico

**Felipe Galván-Magaña**
Instituto Politécnico Nacional, Centro Interdisciplinario de Ciencias Marinas, La Paz, Mexico

**Emiliano García-Rodríguez**
Departamento de Oceanografía Biológica, Centro de Investigación Científica y de Educación Superior de Ensenada (CICESE), Ensenada, Baja California, México

**Mauricio Hoyos-Padilla**
Pelagios-Kakunjá, La Paz, Mexico

**James Ketchum**
Pelagios-Kakunjá, La Paz, Mexico

**A. Peter Klimley**
Biotelemetry Consultants and Contractors, Petaluma, CA, United States

**Shawn Larson**
Seattle Aquarium, Department of Conservation Programs and Partnerships, Seattle, WA, United States

**Dayv Lowry**
Washington Department of Fish and Wildlife, Fish Management Division, Marine Fish Science Unit, Olympia, WA, United States

**Juan Carlos Pérez-Jiménez**
El Colegio de la Frontera Sur (ECOSUR), Lerma, Campeche, México

**Sergio Ramírez-Amaro**
Instituto Español de Oceanografía, Centre Oceanogràfic de les Balears; Laboratori de Genètica, Universitat de les Illes Balears, Palma, Spain

**Emmanuel Rivera-Téllez**
Coordination of Mexican CITES Scientific Authority, National Commission for the Knowledge and Use of Biodiversity (CONABIO), Mexico City, México

**Luz Erandi Saldaña-Ruiz**
Departamento de Oceanografía Biológica, Centro de Investigación Científica y de Educación Superior de Ensenada (CICESE), Ensenada, Baja California, México

**Jonathan Sandoval-Castillo**
Molecular Ecology Lab, College of Science and Engineering, Flinders University, Adelaide, SA, Australia

**Yassir Eden Torres-Rojas**
Instituto de Ecología, Pesquerías y Oceanografía del Golfo de México, Universidad Autónoma de Campeche (EPOMEX-UAC), Campeche, Mexico

**Javier Tovar-Ávila**
Instituto Nacional de Pesca (INAPESCA), Centro Regional de Investigacion Pesquera (CRIP), La Cruz de Huanacaxtle, Nayarit, México

# Contents

# Series Contents for Last Fifteen Years*,†

Volume 44, 2003.
Hirst, A. G., Roff, J. C. and Lampitt, R. S. A synthesis of growth rates in epipelagic invertebrate zooplankton. pp. 3–142.
Boletzky, S. von. Biology of early life stages in cephalopod molluscs. pp. 143–203.
Pittman, S. J. and McAlpine, C. A. Movements of marine fish and decapod crustaceans: process, theory and application. pp. 205–294.
Cutts, C. J. Culture of harpacticoid copepods: potential as live feed for rearing marine fish. pp. 295–315.

Volume 45, 2003.
Cumulative Taxonomic and Subject Index.

Volume 46, 2003.
Gooday, A. J. Benthic foraminifera (Protista) as tools in deep-water palaeoceanography: environmental influences on faunal characteristics. pp. 1–90.
Subramoniam, T. and Gunamalai, V. Breeding biology of the intertidal sand crab, Emerita (Decapoda: Anomura). pp. 91–182.
Coles, S. L. and Brown, B. E. Coral bleaching—capacity for acclimatization and adaptation. pp. 183–223.
Dalsgaard J., St. John M., Kattner G., Müller-Navarra D. and Hagen W. Fatty acid trophic markers in the pelagic marine environment. pp. 225–340.

Volume 47, 2004.
Southward, A. J., Langmead, O., Hardman-Mountford, N. J., Aiken, J., Boalch, G. T., Dando, P. R., Genner, M. J., Joint, I., Kendall, M. A., Halliday, N. C., Harris, R. P., Leaper, R., Mieszkowska, N., Pingree, R. D., Richardson, A. J., Sims, D.W., Smith, T., Walne, A. W. and Hawkins, S. J. Long-term oceanographic and ecological research in the western English Channel. pp. 1–105.
Queiroga, H. and Blanton, J. Interactions between behaviour and physical forcing in the control of horizontal transport of decapod crustacean larvae. pp. 107–214.

*The full list of contents for volumes 1–37 can be found in volume 38
†The full list of contents for volumes 38–43 can be found in volume 78

Braithwaite, R. A. and McEvoy, L. A. Marine biofouling on fish farms and its remediation. pp. 215–252.

Frangoulis, C., Christou, E. D. and Hecq, J. H. Comparison of marine copepod outfluxes: nature, rate, fate and role in the carbon and nitrogen cycles. pp. 253–309.

Volume 48, 2005.

Canfield, D. E., Kristensen, E. and Thamdrup, B. Aquatic Geomicrobiology. pp. 1–599.

Volume 49, 2005.

Bell, J. D., Rothlisberg, P. C., Munro, J. L., Loneragan, N. R., Nash, W. J., Ward, R. D. and Andrew, N. L. Restocking and stock enhancement of marine invertebrate fisheries. pp. 1–358.

Volume 50, 2006.

Lewis, J. B. Biology and ecology of the hydrocoral Millepora on coral reefs. pp. 1–55.

Harborne, A. R., Mumby, P. J., Micheli, F., Perry, C. T., Dahlgren, C. P., Holmes, K. E., and Brumbaugh, D. R. The functional value of Caribbean coral reef, seagrass and mangrove habitats to ecosystem processes. pp. 57–189.

Collins, M. A. and Rodhouse, P. G. K. Southern ocean cephalopods. pp. 191–265.

Tarasov, V. G. Effects of shallow-water hydrothermal venting on biological communities of coastal marine ecosystems of the western Pacific. pp. 267–410.

Volume 51, 2006.

Elena Guijarro Garcia. The fishery for Iceland scallop (Chlamys islandica) in the Northeast Atlantic. pp. 1–55.

Jeffrey, M. Leis. Are larvae of demersal fishes plankton or nekton? pp. 57–141.

John C. Montgomery, Andrew Jeffs, Stephen D. Simpson, Mark Meekan and Chris Tindle. Sound as an orientation cue for the pelagic larvae of reef fishes and decapod crustaceans. pp. 143–196.

Carolin E. Arndt and Kerrie M. Swadling. Crustacea in Arctic and Antarctic sea ice: Distribution, diet and life history strategies. pp. 197–315.

Volume 52, 2007.
Leys, S. P., Mackie, G. O. and Reiswig, H. M. The Biology of Glass Sponges. pp. 1–145.
Garcia E. G. The Northern Shrimp (Pandalus borealis) Offshore Fishery in the Northeast Atlantic. pp. 147–266.
Fraser K. P. P. and Rogers A. D. Protein Metabolism in Marine Animals: The Underlying Mechanism of Growth. pp. 267–362.

Volume 53, 2008.
Dustin J. Marshall and Michael J. Keough. The Evolutionary Ecology of Offspring Size in Marine Invertebrates. pp. 1–60.
Kerry A. Naish, Joseph E. Taylor III, Phillip S. Levin, Thomas P. Quinn, James R. Winton, Daniel Huppert, and Ray Hilborn. An Evaluation of the Effects of Conservation and Fishery Enhancement Hatcheries on Wild Populations of Salmon. pp. 61–194.
Shannon Gowans, Bernd Würsig, and Leszek Karczmarski. The Social Structure and Strategies of Delphinids: Predictions Based on an Ecological Framework. pp. 195–294.

Volume 54, 2008.
Bridget S. Green. Maternal Effects in Fish Populations. pp. 1–105.
Victoria J. Wearmouth and David W. Sims. Sexual Segregation in Marine Fish, Reptiles, Birds and Mammals: Behaviour Patterns, Mechanisms and Conservation Implications. pp. 107–170.
David W. Sims. Sieving a Living: A Review of the Biology, Ecology and Conservation Status of the Plankton-Feeding Basking Shark *Cetorhinus Maximus*. pp. 171–220.
Charles H. Peterson, Kenneth W. Able, Christin Frieswyk DeJong, Michael F. Piehler, Charles A. Simenstad, and Joy B. Zedler. Practical Proxies for Tidal Marsh Ecosystem Services: Application to Injury and Restoration. pp. 221–266.

Volume 55, 2008.
Annie Mercier and Jean-Francois Hamel. Introduction. pp. 1–6.
Annie Mercier and Jean-Francois Hamel. Gametogenesis. pp. 7–72.
Annie Mercier and Jean-Francois Hamel. Spawning. pp. 73–168.
Annie Mercier and Jean-Francois Hamel. Discussion. pp. 169–194.

Volume 56, 2009.
Philip C. Reid, Astrid C. Fischer, Emily Lewis-Brown, Michael P. Meredith, Mike Sparrow, Andreas J. Andersson, Avan Antia, Nicholas R. Bates, Ulrich Bathmann, Gregory Beaugrand, Holger Brix, Stephen Dye, Martin Edwards, Tore Furevik, Reidun Gangst, Hjalmar Hatun, Russell R. Hopcroft, Mike Kendall, Sabine Kasten, Ralph Keeling, Corinne Le Quere, Fred T. Mackenzie, Gill Malin, Cecilie Mauritzen, Jon Olafsson, Charlie Paull, Eric Rignot, Koji Shimada, Meike Vogt, Craig Wallace, Zhaomin Wang and Richard Washington. Impacts of the Oceans on Climate Change. pp. 1–150.
Elvira S. Poloczanska, Colin J. Limpus and Graeme C. Hays. Vulnerability of Marine Turtles to Climate Change. pp. 151–212.
Nova Mieszkowska, Martin J. Genner, Stephen J. Hawkins and David W. Sims. Effects of Climate Change and Commercial Fishing on Atlantic Cod *Gadus morhua*. pp. 213–274.
Iain C. Field, Mark G. Meekan, Rik C. Buckworth and Corey J. A. Bradshaw. Susceptibility of Sharks, Rays and Chimaeras to Global Extinction. pp. 275–364.
Milagros Penela-Arenaz, Juan Bellas and Elsa Vazquez. Effects of the *Prestige* Oil Spill on the Biota of NW Spain: 5 Years of Learning. pp. 365–396.

Volume 57, 2010.
Geraint A. Tarling, Natalie S. Ensor, Torsten Fregin, William P. Good-all-Copestake and Peter Fretwell. An Introduction to the Biology of Northern Krill (*Meganyctiphanes norvegica* Sars). pp. 1–40.
Tomaso Patarnello, Chiara Papetti and Lorenzo Zane. Genetics of Northern Krill (*Meganyctiphanes norvegica* Sars). pp. 41–58.
Geraint A. Tarling. Population Dynamics of Northern Krill (*Meganyctiphanes norvegica* Sars). pp. 59–90.
John I. Spicer and Reinhard Saborowski. Physiology and Metabolism of Northern Krill (*Meganyctiphanes norvegica* Sars). pp. 91–126.
Katrin Schmidt. Food and Feeding in Northern Krill (*Meganyctiphanes norvegica* Sars). pp. 127–172.
Friedrich Buchholz and Cornelia Buchholz. Growth and Moulting in Northern Krill (*Meganyctiphanes norvegica* Sars). pp. 173–198.
Janine Cuzin-Roudy. Reproduction in Northern Krill. pp. 199–230.
Edward Gaten, Konrad Wiese and Magnus L. Johnson. Laboratory-Based Observations of Behaviour in Northern Krill (*Meganyctiphanes norvegica* Sars). pp. 231–254.

Stein Kaartvedt. Diel Vertical Migration Behaviour of the Northern Krill (*Meganyctiphanes norvegica* Sars). pp. 255–276.
Yvan Simard and Michel Harvey. Predation on Northern Krill (*Meganyctiphanes norvegica* Sars). pp. 277–306.

Volume 58, 2010.
A. G. Glover, A. J. Gooday, D. M. Bailey, D. S. M. Billett, P. Chevaldonné, A. Colaço, J. Copley, D. Cuvelier, D. Desbruyères, V. Kalogeropoulou, M. Klages, N. Lampadariou, C. Lejeusne, N. C. Mestre, G. L. J. Paterson, T. Perez, H. Ruhl, J. Sarrazin, T. Soltwedel, E. H. Soto, S. Thatje, A. Tselepides, S. Van Gaever, and A. Vanreusel. Temporal Change in Deep-Sea Benthic Ecosystems: A Review of the Evidence From Recent Time-Series Studies. pp. 1–96.
Hilario Murua. The Biology and Fisheries of European Hake, *Merluccius merluccius*, in the North-East Atlantic. pp. 97–154.
Jacopo Aguzzi and Joan B. Company. Chronobiology of Deep-Water Decapod Crustaceans on Continental Margins. pp. 155–226.
Martin A. Collins, Paul Brickle, Judith Brown, and Mark Belchier. The Patagonian Toothfish: Biology, Ecology and Fishery. pp. 227–300.

Volume 59, 2011.
Charles W. Walker, Rebecca J. Van Beneden, Annette F. Muttray, S. Anne Böttger, Melissa L. Kelley, Abraham E. Tucker, and W. Kelley Thomas. p53 Superfamily Proteins in Marine Bivalve Cancer and Stress Biology. pp 1–36.
Martin Wahl, Veijo Jormalainen, Britas Klemens Eriksson, James A. Coyer, Markus Molis, Hendrik Schubert, Megan Dethier, Anneli Ehlers, Rolf Karez, Inken Kruse, Mark Lenz, Gareth Pearson, Sven Rohde, Sofia A. Wikström, and Jeanine L. Olsen. Stress Ecology in *Fucus*: Abiotic, Biotic and Genetic Interactions. pp. 37–106.
Steven R. Dudgeon and Janet E. Kübler. Hydrozoans and the Shape of Things to Come. pp. 107–144.
Miles Lamare, David Burritt, and Kathryn Lister. Ultraviolet Radiation and Echinoderms: Past, Present and Future Perspectives. pp. 145–187.

Volume 60, 2011.
Tatiana A. Rynearson and Brian Palenik. Learning to Read the Oceans: Genomics of Marine Phytoplankton. pp. 1–40.
Les Watling, Scott C. France, Eric Pante and Anne Simpson. Biology of Deep-Water Octocorals. pp. 41–122.

Cristián J. Monaco and Brian Helmuth. Tipping Points, Thresholds and the Keystone Role of Physiology in Marine Climate Change Research. pp. 123–160.
David A. Ritz, Alistair J. Hobday, John C. Montgomery and Ashley J.W. Ward. Social Aggregation in the Pelagic Zone with Special Reference to Fish and Invertebrates. pp. 161–228.

Volume 61, 2012.
Gert Wörheide, Martin Dohrmann, Dirk Erpenbeck, Claire Larroux, Manuel Maldonado, Oliver Voigt, Carole Borchiellini and Denis Lavrov. Deep Phylogeny and Evolution of Sponges (Phylum Porifera). pp. 1–78.
Paco Cárdenas, Thierry Pérez and Nicole Boury-Esnault. Sponge Systematics Facing New Challenges. pp. 79–210.
Klaus Rützler. The Role of Sponges in the Mesoamerican Barrier-Reef Ecosystem, Belize. pp. 211–272.
Janie Wulff. Ecological Interactions and the Distribution, Abundance, and Diversity of Sponges. pp. 273–344.
Maria J. Uriz and Xavier Turon. Sponge Ecology in the Molecular Era. pp. 345–410.

Volume 62, 2012.
Sally P. Leys and April Hill. The Physiology and Molecular Biology of Sponge Tissues. pp. 1–56.
Robert W. Thacker and Christopher J. Freeman. Sponge–Microbe Symbioses: Recent Advances and New Directions. pp. 57–112.
Manuel Maldonado, Marta Ribes and Fleur C. van Duyl. Nutrient Fluxes Through Sponges: Biology, Budgets, and Ecological Implications. pp. 113–182.
Grégory Genta-Jouve and Olivier P. Thomas. Sponge Chemical Diversity: From Biosynthetic Pathways to Ecological Roles. pp. 183–230.
Xiaohong Wang, Heinz C. Schröder, Matthias Wiens, Ute Schloßmacher and Werner E. G. Müller. Biosilica: Molecular Biology, Biochemistry and Function in Demosponges as well as its Applied Aspects for Tissue Engineering. pp. 231–272.
Klaske J. Schippers, Detmer Sipkema, Ronald Osinga, Hauke Smidt, Shirley A. Pomponi, Dirk E. Martens and René H. Wijffels. Cultivation of Sponges, Sponge Cells and Symbionts: Achievements and Future Prospects. pp. 273–338.

Volume 63, 2012.

Michael Stat, Andrew C. Baker, David G. Bourne, Adrienne M. S. Correa, Zac Forsman, Megan J. Huggett, Xavier Pochon, Derek Skillings, Robert J. Toonen, Madeleine J. H. van Oppen, and Ruth D. Gates. Molecular Delineation of Species in the Coral Holobiont. pp. 1–66.

Daniel Wagner, Daniel G. Luck, and Robert J. Toonen. The Biology and Ecology of Black Corals (Cnidaria: Anthozoa: Hexacorallia: Antipatharia). pp. 67–132.

Cathy H. Lucas, William M. Graham, and Chad Widmer. Jellyfish Life Histories: Role of Polyps in Forming and Maintaining Scyphomedusa Populations. pp. 133–196.

T. Aran Mooney, Maya Yamato, and Brian K. Branstetter. Hearing in Cetaceans: From Natural History to Experimental Biology. pp. 197–246.

Volume 64, 2013.

Dale Tshudy. Systematics and Position of *Nephrops* Among the Lobsters. pp. 1–26.

Mark P. Johnson, Colm Lordan, and Anne Marie Power. Habitat and Ecology of *Nephrops norvegicus*. pp. 27–64.

Emi Katoh, Valerio Sbragaglia, Jacopo Aguzzi, and Thomas Breithaupt. Sensory Biology and Behaviour of *Nephrops norvegicus*. pp. 65–106.

Edward Gaten, Steve Moss, and Magnus L. Johnson. The Reniform Reflecting Superposition Compound Eyes of *Nephrops norvegicus*: Optics, Susceptibility to Light-Induced Damage, Electrophysiology and a Ray Tracing Model. pp. 107–148.

Susanne P. Eriksson, Bodil Hernroth, and Susanne P. Baden. Stress Biology and Immunology in *Nephrops norvegicus*. pp. 149–200.

Adam Powell and Susanne P. Eriksson. Reproduction: Life Cycle, Larvae and Larviculture. pp. 201–246.

Anette Ungfors, Ewen Bell, Magnus L. Johnson, Daniel Cowing, Nicola C. Dobson, Ralf Bublitz, and Jane Sandell. *Nephrops* Fisheries in European Waters. pp. 247–314.

Volume 65, 2013.

Isobel S.M. Bloor, Martin J. Attrill, and Emma L. Jackson. A Review of the Factors Influencing Spawning, Early Life Stage Survival and Recruitment Variability in the Common Cuttlefish (*Sepia officinalis*). pp. 1–66.

Dianna K. Padilla and Monique M. Savedo. A Systematic Review of Phenotypic Plasticity in Marine Invertebrate and Plant Systems. pp. 67–120.

Leif K. Rasmuson. The Biology, Ecology and Fishery of the Dungeness crab, *Cancer magister*. pp. 121–174.

Volume 66, 2013.
Lisa-ann Gershwin, Anthony J. Richardson, Kenneth D. Winkel, Peter J. Fenner, John Lippmann, Russell Hore, Griselda Avila-Soria, David Brewer, Rudy J. Kloser, Andy Steven, and Scott Condie. Biology and Ecology of Irukandji Jellyfish (Cnidaria: Cubozoa). pp. 1–86.
April M. H. Blakeslee, Amy E. Fowler, and Carolyn L. Keogh. Marine Invasions and Parasite Escape: Updates and New Perspectives. pp. 87–170.
Michael P. Russell. Echinoderm Responses to Variation in Salinity. pp. 171–212.
Daniela M. Ceccarelli, A. David McKinnon, Serge Andréfouët, Valerie Allain, Jock Young, Daniel C. Gledhill, Adrian Flynn, Nicholas J. Bax, Robin Beaman, Philippe Borsa, Richard Brinkman, Rodrigo H. Bustamante, Robert Campbell, Mike Cappo, Sophie Cravatte, Stéphanie D'Agata, Catherine M. Dichmont, Piers K. Dunstan, Cécile Dupouy, Graham Edgar, Richard Farman, Miles Furnas, Claire Garrigue, Trevor Hutton, Michel Kulbicki, Yves Letourneur, Dhugal Lindsay, Christophe Menkes, David Mouillot, Valeriano Parravicini, Claude Payri, Bernard Pelletier, Bertrand Richer de Forges, Ken Ridgway, Martine Rodier, Sarah Samadi, David Schoeman, Tim Skewes, Steven Swearer, Laurent Vigliola, Laurent Wantiez, Alan Williams, Ashley Williams, and Anthony J. Richardson. The Coral Sea: Physical Environment, Ecosystem Status and Biodiversity Assets. pp. 213–290.

Volume 67, 2014.
Erica A.G. Vidal, Roger Villanueva, José P. Andrade, Ian G. Gleadall, José Iglesias, Noussithé Koueta, Carlos Rosas, Susumu Segawa, Bret Grasse, Rita M. Franco-Santos, Caroline B. Albertin, Claudia Caamal-Monsreal, Maria E. Chimal, Eric Edsinger-Gonzales, Pedro Gallardo, Charles Le Pabic, Cristina Pascual, Katina Roumbedakis, and James Wood. Cephalopod Culture: Current Status of Main Biological Models and Research Priorities. pp. 1–98.
Paul G.K. Rodhouse, Graham J. Pierce, Owen C. Nichols, Warwick H.H. Sauer, Alexander I. Arkhipkin, Vladimir V. Laptikhovsky, Marek R. Lipiński, Jorge E. Ramos, Michaël Gras, Hideaki Kidokoro, Kazuhiro Sadayasu, João Pereira, Evgenia Lefkaditou, Cristina Pita, Maria Gasalla, Manuel Haimovici, Mitsuo Sakai, and Nicola Downey. Environmental Effects on Cephalopod Population Dynamics: Implications for Management of Fisheries. pp. 99–234.

Henk-Jan T. Hoving, José A.A. Perez, Kathrin Bolstad, Heather Braid, Aaron B. Evans, Dirk Fuchs, Heather Judkins, Jesse T. Kelly, José E.A.R. Marian, Ryuta Nakajima, Uwe Piatkowski, Amanda Reid, Michael Vecchione, and José C.C. Xavier. The Study of Deep-Sea Cephalopods. pp. 235–362.

Jean-Paul Robin, Michael Roberts, Lou Zeidberg, Isobel Bloor, Almendra Rodriguez, Felipe Briceño, Nicola Downey, Maite Mascaró, Mike Navarro, Angel Guerra, Jennifer Hofmeister, Diogo D. Barcellos, Silvia A.P. Lourenço, Clyde F.E. Roper, Natalie A. Moltschaniwskyj, Corey P. Green, and Jennifer Mather. Transitions During Cephalopod Life History: The Role of Habitat, Environment, Functional Morphology and Behaviour. pp. 363–440.

Volume 68, 2014.

Paul K.S. Shin, Siu Gin Cheung, Tsui Yun Tsang, and Ho Yin Wai. Ecology of Artificial Reefs in the Subtropics. pp. 1–64.

Hrafnkell Eiríksson. Reproductive Biology of Female Norway Lobster, *Nephrops norvegicus* (Linnaeus, 1758) Leach, in Icelandic Waters During the Period 1960–2010: Comparative Overview of Distribution Areas in the Northeast Atlantic and the Mediterranean. pp. 65–210.

Volume 69, 2014.

Ray Hilborn. Introduction to Marine Managed Areas. pp. 1–14.

Philip N. Trathan, Martin A. Collins, Susie M. Grant, Mark Belchier, David K.A. Barnes, Judith Brown, and Iain J. Staniland. The South Georgia and the South Sandwich Islands MPA: Protecting A Biodiverse Oceanic Island Chain Situated in the Flow of the Antarctic Circumpolar Current. pp. 15–78.

Richard P. Dunne, Nicholas V.C. Polunin, Peter H. Sand, and Magnus L. Johnson. The Creation of the Chagos Marine Protected Area: A Fisheries Perspective. pp. 79–128.

Michelle T. Schärer-Umpierre, Daniel Mateos-Molina, Richard Appeldoorn, Ivonne Bejarano, Edwin A. Hernández-Delgado, Richard S. Nemeth, Michael I. Nemeth, Manuel Valdés-Pizzini, and Tyler B. Smith. Marine Managed Areas and Associated Fisheries in the US Caribbean. pp. 129–152.

Alan M. Friedlander, Kostantinos A. Stamoulis, John N. Kittinger, Jeffrey C. Drazen, and Brian N. Tissot. Understanding the Scale of Marine Protection in Hawai'i: From Community-Based Management to the Remote Northwestern Hawaiian Islands. pp. 153–204.

Louis W. Botsford, J. Wilson White, Mark H. Carr, and Jennifer E. Caselle. Marine Protected Area Networks in California, USA. pp. 205–252.
Bob Kearney and Graham Farebrother. Inadequate Evaluation and Management of Threats in Australia's Marine Parks, Including the Great Barrier Reef, Misdirect Marine Conservation. pp. 253–288.
Randi Rotjan, Regen Jamieson, Ben Carr, Les Kaufman, Sangeeta Mangubhai, David Obura, Ray Pierce, Betarim Rimon, Bud Ris, Stuart Sandin, Peter Shelley, U. Rashid Sumaila, Sue Taei, Heather Tausig, Tukabu Teroroko, Simon Thorrold, Brooke Wikgren, Teuea Toatu, and Greg Stone. Establishment, Management, and Maintenance of the Phoenix Islands Protected Area. pp. 289–324.
Alex J. Caveen, Clare Fitzsimmons, Margherita Pieraccini, Euan Dunn, Christopher J. Sweeting, Magnus L. Johnson, Helen Bloomfield, Estelle V. Jones, Paula Lightfoot, Tim S. Gray, Selina M. Stead, and Nicholas V. C. Polunin. Diverging Strategies to Planning an Ecologically Coherent Network of MPAs in the North Sea: The Roles of Advocacy, Evidence and Pragmatism in the Face of Uncertaintya. pp. 325–370.
Carlo Pipitone, Fabio Badalamenti, Tomás Vega Fernández, and Giovanni D'Anna. Spatial Management of Fisheries in the Mediterranean Sea: Problematic Issues and a Few Success Stories. pp. 371–402.

Volume 70, 2015.
Alex D. Rogers, Christopher Yesson, and Pippa Gravestock. A Biophysical and Economic Profile of South Georgia and the South Sandwich Islands as Potential Large-Scale Antarctic Protected Areas. pp. 1–286.

Volume 71, 2015.
Ricardo Calado and Miguel Costa Leal. Trophic Ecology of Benthic Marine Invertebrates with Bi-Phasic Life Cycles: What Are We Still Missing? pp. 1–70.
Jesse M.A. van der Grient and Alex D. Rogers. Body Size Versus Depth: Regional and Taxonomical Variation in Deep-Sea Meio- and Macro-faunal Organisms. pp. 71–108.
Lorena Basso, Maite Vázquez-Luis, José R. García-March, Salud Deudero, Elvira Alvarez, Nardo Vicente, Carlos M. Duarte, and Iris E. Hendriks. The Pen Shell, *Pinna nobilis*: A Review of Population Status and Recommended Research Priorities in the Mediterranean Sea. pp. 109–160.

Volume 72, 2015.
Thomas A. Jefferson and Barbara E. Curry. Humpback Dolphins: A Brief Introduction to the Genus *Sousa*. pp. 1–16.
Sarah Piwetz, David Lundquist, and Bernd Würsig. Humpback Dolphin (Genus *Sousa*) Behavioural Responses to Human Activities. pp. 17–46.
Tim Collins. Re-assessment of the Conservation Status of the Atlantic Humpback Dolphin, *Sousa teuszii* (Kükenthal, 1892) Using the IUCN Red List Criteria. pp. 47–78.
Caroline R. Weir and Tim Collins. A Review of the Geographical Distribution and Habitat of the Atlantic Humpback Dolphin (*Sousa teuszii*). pp. 79–118.
Gill T Braulik, Ken Findlay, Salvatore Cerchio, and Robert Baldwin. Assessment of the Conservation Status of the Indian Ocean Humpback Dolphin (*Sousa plumbea*) Using the IUCN Red List Criteria. pp. 119–142.
Stephanie Plön, Victor G. Cockcroft, and William P. Froneman. The Natural History and Conservation of Indian Ocean Humpback Dolphins (*Sousa plumbea*) in South African Waters. pp. 143–162.
Salvatore Cerchio, Norbert Andrianarivelo, and Boris Andrianantenaina. Ecology and Conservation Status of Indian Ocean Humpback Dolphins (*Sousa plumbea*) in Madagascar. pp. 163–200.
Muhammad Shoaib Kiani and Koen Van Waerebeek. A Review of the Status of the Indian Ocean Humpback Dolphin *Sousa plumbea* in Pakistan. pp. 201–228.
Dipani Sutaria, Divya Panicker, Ketki Jog, Mihir Sule, Rahul Muralidharan, and Isha Bopardikar. Humpback Dolphins (Genus *Sousa*) in India: An Overview of Status and Conservation Issues. pp. 229–256.

Volume 73, 2016.
Thomas A. Jefferson and Brian D. Smith. Re-assessment of the Conservation Status of the Indo-Pacific Humpback Dolphin (*Sousa chinensis*) Using the IUCN Red List Criteria. pp. 1–26.
Leszek Karczmarski, Shiang-Lin Huang, Carmen K. M. Or, Duan Gui, Stephen C. Y. Chan, Wenzhi Lin, Lindsay Porter, Wai-Ho Wong, Ruiqiang Zheng, Yuen-Wa Ho, Scott Y. S. Chui, Angelico Jose C. Tiongson, Yaqian Mo, Wei-Lun Chang, John H. W. Kwok, Ricky W. K. Tang, Andy T. L. Lee, Sze-Wing Yiu, Mark Keith, Glenn Gailey, and Yuping Wu. Humpback Dolphins in Hong Kong and the Pearl River Delta: Status, Threats and Conservation Challenges. pp. 27–64.

Bernd Würsig, E.C.M. Parsons, Sarah Piwetz, and Lindsay Porter. The Behavioural Ecology of Indo-Pacific Humpback Dolphins in Hong Kong. pp. 65–90.

John Y. Wang, Kimberly N. Riehl, Michelle N. Klein, Shiva Javdan, Jordan M. Hoffman, Sarah Z. Dungan, Lauren E. Dares, and Claryana Araújo-Wang. Biology and Conservation of the Taiwanese Humpback Dolphin, *Sousa chinensis taiwanensis*. pp. 91–118.

Bingyao Chen, Xinrong Xu, Thomas A. Jefferson, Paula A. Olson, Qiurong Qin, Hongke Zhang, Liwen He, and Guang Yang. Conservation Status of the Indo-Pacific Humpback Dolphin (*Sousa chinensis*) in the Northern Beibu Gulf, China. pp. 119–140.

Gianna Minton, Anna Norliza Zulkifli Poh, Cindy Peter, Lindsay Porter, and Danielle Kreb. Indo-Pacific Humpback Dolphins in Borneo: A Review of Current Knowledge with Emphasis on Sarawak. pp. 141–156.

Guido J. Parra and Daniele Cagnazzi. Conservation Status of the Australian Humpback Dolphin (*Sousa sahulensis*) Using the IUCN Red List Criteria. pp. 157–192.

Daniella M. Hanf, Tim Hunt, and Guido J. Parra. Humpback Dolphins of Western Australia: A Review of Current Knowledge and Recommendations for Future Management. pp. 193–218.

Isabel Beasley, Maria Jedensjö, Gede Mahendra Wijaya, Jim Anamiato, Benjamin Kahn, and Danielle Kreb. Observations on Australian Humpback Dolphins (*Sousa sahulensis*) in Waters of the Pacific Islands and New Guinea. pp. 219–272.

Alexander M. Brown, Lars Bejder, Guido J. Parra, Daniele Cagnazzi, Tim Hunt, Jennifer L. Smith, and Simon J. Allen. Sexual Dimorphism and Geographic Variation in Dorsal Fin Features of Australian Humpback Dolphins, *Sousa sahulensis*. pp. 273–314.

Volume 74, 2016.

J. Salinger, A.J. Hobday, R.J. Matear, T.J. O'Kane, J.S. Risbey, P. Dunstan, J.P. Eveson, E.A. Fulton, M. Feng, É.E. Plagányi, E.S. Poloczanska, A.G. Marshall, and P.A. Thompson. Decadal-Scale Forecasting of Climate Drivers for Marine Applications. pp. 1–68.

S.A. Foo and M. Byrne. Acclimatization and Adaptive Capacity of Marine Species in a Changing Ocean. pp. 69–116.

N.D. Gallo and L.A. Levin. Fish Ecology and Evolution in the World's Oxygen Minimum Zones and Implications of Ocean Deoxygenation. pp. 117–198.

R.J. Olson, J.W. Young, F. Ménard, M. Potier, V. Allain, N. Goñi, J.M. Logan, and F. Galván-Magaña. Bioenergetics, Trophic Ecology, and Niche Separation of Tunas. pp. 199–344.

Volume 75, 2016.

G. Notarbartolo di Sciara. Marine Mammals in the Mediterranean Sea: An Overview. pp. 1–36.

L. Rendell and A. Frantzis. Mediterranean Sperm Whales, *Physeter macrocephalus*: The Precarious State of a Lost Tribe. pp. 37–74.

G. Notarbartolo di Sciara, M. Castellote, J.-N. Druon, and S. Panigada. Fin Whales, *Balaenoptera physalus*: At Home in a Changing Mediterranean Sea?. pp. 75–102.

M. Podestà, A. Azzellino, A. Cañadas, A. Frantzis, A. Moulins, M. Rosso, P. Tepsich, and C. Lanfredi. Cuvier's Beaked Whale, *Ziphius cavirostris*, Distribution and Occurrence in the Mediterranean Sea: High-Use Areas and Conservation Threats. pp. 103–140.

R. Esteban, P. Verborgh, P. Gauffier, D. Alarcón, J.M. Salazar-Sierra, J. Giménez, A.D. Foote, and R. de Stephanis. Conservation Status of Killer Whales, *Orcinus orca*, in the Strait of Gibraltar. pp. 141–172.

P. Verborgh, P. Gauffier, R. Esteban, J. Giménez, A. Cañadas, J.M. Salazar-Sierra, and R. de Stephanis. Conservation Status of Long-Finned Pilot Whales, *Globicephala melas*, in the Mediterranean Sea. pp. 173–204.

A. Azzellino, S. Airoldi, S. Gaspari, C. Lanfredi, A. Moulins, M. Podestà, M. Rosso, and P. Tepsich. Risso's Dolphin, *Grampus griseus*, in the Western Ligurian Sea: Trends in Population Size and Habitat Use. pp. 205–232.

D. Kerem, O. Goffman, M. Elasar, N. Hadar, A. Scheinin, and T. Lewis. The Rough-Toothed Dolphin, *Steno bredanensis*, in the Eastern Mediterranean Sea: A Relict Population?. pp. 233–258.

J. Gonzalvo, G. Lauriano, P.S. Hammond, K.A. Viaud-Martinez, M.C. Fossi, A. Natoli, and L. Marsili. The Gulf of Ambracia's Common Bottlenose Dolphins, *Tursiops truncatus*: A Highly Dense and yet Threatened Population. pp. 259–296.

G. Bearzi, S. Bonizzoni, N.L. Santostasi, N.B. Furey, L. Eddy, V.D. Valavanis, and O. Gimenez. Dolphins in a Scaled-Down Mediterranean: The Gulf of Corinth's Odontocetes. pp. 297–332.

M.C. Fontaine. Harbour Porpoises, *Phocoena phocoena*, in the Mediterranean Sea and Adjacent Regions: Biogeographic Relicts of the Last Glacial Period. pp. 333–358.

G. Notarbartolo di Sciara and S. Kotomatas. Are Mediterranean Monk Seals, *Monachus monachus*, Being Left to Save Themselves from Extinction?. pp. 359–386.
T. Scovazzi. The International Legal Framework for Marine Mammal Conservation in the Mediterranean Sea. pp. 387–416.

Volume 76, 2017.
K.S. Meyer. Islands in a Sea of Mud: Insights From Terrestrial Island Theory for Community Assembly on Insular Marine Substrata. pp. 1–40.
E.M. Montgomery, J.-F. Hamel, and A. Mercier. Patterns and Drivers of Egg Pigment Intensity and Colour Diversity in the Ocean: A Meta-Analysis of Phylum Echinodermata. pp. 41–104.
F. Espinosa and G.A. Rivera-Ingraham. Biological Conservation of Giant Limpets: The Implications of Large Size. pp. 105–156.
L. Yebra, T. Kobari, A.R. Sastri, F. Gusmão, and S. Hernández-León. Advances in Biochemical Indices of Zooplankton Production. pp. 157–240.

Volume 77, 2017.
Dayv Lowry and Shawn E. Larson. Introduction to Northeast Pacific Shark Biology, Ecology, and Conservation. pp. 1–8.
David A. Ebert, Jennifer S. Bigman, and Julia M. Lawson. Biodiversity, Life History, and Conservation of Northeastern Pacific Chondrichthyans pp. 9–78.
Shawn E. Larson, Toby S. Daly-Engel, and Nicole M. Phillips. Review of Current Conservation Genetic Analyses of Northeast Pacific Sharks. pp. 79–110.
Joseph J. Bizzarro, Aaron B. Carlisle, Wade D. Smith, and Enric Cortés. Diet Composition and Trophic Ecology of Northeast Pacific Ocean Sharks. pp. 111–148.
Jonathan C.P. Reum, Gregory D. Williams, and Chris J. Harvey. Stable Isotope Applications for Understanding Shark Ecology in the Northeast Pacific Ocean. pp. 149–178.
Mary E. Matta, Cindy A. Tribuzio, David A. Ebert, Kenneth J. Goldman, and Christopher M. Gburski. Age and Growth of Elasmobranchs and Applications to Fisheries Management and Conservation in the Northeast Pacific Ocean. pp. 179–220.

Volume 78, 2017.
Shawn E. Larson and Dayv Lowry. Introduction to Northeast Pacific Shark Biology, Research, and Conservation, Part B. pp. 1–8.
Jackie King, Gordon A. McFarlane, Vladlena Gertseva, Jason Gasper, Sean Matson, and Cindy A. Tribuzio. Shark Interactions With Directed and Incidental Fisheries in the Northeast Pacific Ocean: Historic and Current Encounters, and Challenges for Shark Conservation. pp. 9–44.
Dovi Kacev, Timothy J. Sippel, Michael J. Kinney, Sebastián A. Pardo, and Christopher G. Mull. An Introduction to Modelling Abundance and Life History Parameters in Shark Populations. pp. 45–88.
Michael Grassmann, Bryan McNeil, and Jim Wharton. Sharks in Captivity: The Role of Husbandry, Breeding, Education, and Citizen Science in Shark Conservation. pp. 89–120.
Peter A. Mieras, Chris Harvey-Clark, Michael Bear, Gina Hodgin, and Boone Hodgin. The Economy of Shark Conservation in the Northeast Pacific: The Role of Ecotourism and Citizen Science. pp. 121–154.
Dayv Lowry. Conclusions: The Future of Shark Management and Conservation in the Northeast Pacific Ocean. pp. 155–164.

Volume 79, 2018.
Patricia L.M. Lee, Gail Schofield, Rebecca I. Haughey, Antonios D. Mazaris, and Graeme C. Hays. A Review of Patterns of Multiple Paternity Across Sea Turtle Rookeries. pp. 1–32.
Johanne Vad, Georgios Kazanidis, Lea-Anne Henry, Daniel O.B. Jones, Ole S. Tendal, Sabine Christiansen, Theodore B. Henry, and J. Murray Roberts. Potential Impacts of Offshore Oil and Gas Activities on Deep-Sea Sponges and the Habitats They Form. pp. 33–60.
Gianmarco Ingrosso, Marco Abbiati, Fabio Badalamenti, Giorgio Bavestrello, Genuario Belmonte, Rita Cannas, Lisandro Benedetti-Cecchi, Marco Bertolino, Stanislao Bevilacqua, Carlo Nike Bianchi, Marzia Bo, Elisa Boscari, Frine Cardone, Riccardo Cattaneo-Vietti, Alessandro Cau, Carlo Cerrano, Renato Chemello, Giovanni Chimienti, Leonardo Congiu, Giuseppe Corriero, Federica Costantini, Francesco De Leo, Luigia Donnarumma, Annalisa Falace, Simonetta Fraschetti, Adriana Giangrande, Maria Flavia Gravina, Giuseppe Guarnieri, Francesco Mastrototaro, Marco Milazzo, Carla Morri, Luigi Musco, Laura Pezzolesi, Stefano Piraino, Fiorella Prada, Massimo Ponti, Fabio Rindi, Giovanni Fulvio Russo, Roberto Sandulli, Adriana Villamor, Lorenzo Zane, and Ferdinando Boero. Mediterranean Bioconstructions Along the Italian Coast. pp. 61–136.
Alex D. Rogers. The Biology of Seamounts: 25 Years on. pp. 137–224.

Volume 80, 2018.
Michael H. Schleyer and Sean N. Porter. Drivers of Soft and Stony Coral Community Distribution on the High-Latitude Coral Reefs of South Africa. pp. 1–56.
Sara González-Delgado and José Carlos Hernández. The Importance of Natural Acidified Systems in the Study of Ocean Acidification: What Have We Learned? pp. 57–100.

Volume 81, 2018.
Melissa C. Cook, Adam May, Lucas Kohl, Geert Van Biesen, Christopher C. Parrish, and Penny L. Morrill. The Potential Impact of Hydrocarbons on Mussels in Port au Port Bay, Newfoundland pp. 1–22.
Yunwen Tao, Bing Chen, Baiyu (Helen) Zhang, Zhiwen (Joy) Zhu, and Qinhong Cai. Occurrence, Impact, Analysis and Treatment of Metformin and Guanylurea in Coastal Aquatic Environments of Canada, USA and Europe pp. 23–58.
Gong Zhang, Chun Yang, Mariam Serhan, Graeme Koivu, Zeyu Yang, Bruce Hollebone, Patrick Lambert, and Carl E. Brown. Characterization of Nitrogen-Containing Polycyclic Aromatic Heterocycles in Crude Oils and Refined Petroleum Products. pp. 59–96.
Arpana Rani Datta, Qiao Kang, Bing Chen, and Xudong Ye. Fate and Transport Modelling of Emerging Pollutants from Watersheds to Oceans: A Review. pp. 97–128.
Meijin Du, Wenwen Gu, Xixi Li, Fuqiang Fan, and Yu Li. Modification of Hexachlorobenzene to Molecules with Lower Long-Range Transport Potentials Using 3D-QSAR Models with a Full Factor Experimental Design pp. 129–166.
Weiyun Lin, Xixi Li, Min Yang, Kenneth Lee, Bing Chen, and Baiyu (Helen) Zhang. Brominated Flame Retardants, Microplastics, and Biocides in the Marine Environment: Recent Updates of Occurrence, Analysis, and Impacts. pp. 167–212.
Xixi Li, Zhenhua Chu, Jiawen Yang, Minghao Li, Meijin Du, Xiaohui Zhao, Zhiwen (Joy) Zhu, and Yu Li. Synthetic Musks: A Class of Commercial Fragrance Additives in Personal Care Products (PCPs) Causing Concern as Emerging Contaminants. pp. 213–280.

Volume 82, 2019.
Junaid S. Khan, Jennifer F. Provencher, Mark R. Forbes, Mark L. Mallory, Camille Lebarbenchon, and Karen D. McCoy. Parasites of seabirds: A survey of effects and ecological implications. pp. 1–50.

Charlotte Havermans, Holger Auel, Wilhelm Hagen, Christoph Held, Natalie S. Ensor, and Geraint A. Tarling. Predatory zooplankton on the move: *Themisto* amphipods in high-latitude marine pelagic food webs. pp. 51–92.
Angel Borja, Guillem Chust, and Iñigo Muxika. Forever young: The successful story of a marine biotic index. pp. 93–128.
Ferdinando Boero, Francesco De Leo, Simonetta Fraschetti, and Gianmarco Ingrosso. The Cells of Ecosystem Functioning: Towards a holistic vision of marine space. pp. 129–154.

CHAPTER ONE

# Introduction: The sharks of Pacific Mexico and their conservation: Why should we care?

**Dayv Lowry[a,*], Shawn Larson[b]**

[a]Washington Department of Fish and Wildlife, Fish Management Division, Marine Fish Science Unit, Olympia, WA, United States
[b]Seattle Aquarium, Department of Conservation Programs and Partnerships, Seattle, WA, United States
*Corresponding author: e-mail address: dayv.lowry@dfw.wa.gov

## Contents



## Abstract

Sharks exhibit a broad range of body forms, occupy diverse ecological niches, and are found in every ocean on earth. They have roles as both predator and prey and have an evolutionary history far longer than that of human beings. As long as humans have interacted with the ocean they have interacted with sharks and, as a result, sharks have featured in the mythology, history, and fisheries of diverse cultures around the world. Only in the last 100 years or less, however, have humans begun to study and fully realize the susceptibility of these animals to intensive fishing pressure, ecological degradation, and other anthropogenic factors. This volume highlights the biodiversity and biological attributes of, and conservation efforts targeted at, populations of sharks that reside in the Northeast Pacific Ocean (NEP) bordering the west coast of Mexico. Together with a pair of companion volumes published in 2017 that focused on the NEP adjacent to the United States and Canada, this volume provide readers a detailed backdrop of shark biology and conservation in the Mexican waters of the Northeast Pacific against which to consider their own actions, and those of resource managers, academics, and educators, as they relate to the long-term conservation of sharks and their relatives.

## 1. Introduction

Sharks and humans have interacted with one another for millennia, and the tone and tenor of these interactions have differed substantially as a result of geography, ecology, culture, and technology. Among other

*Advances in Marine Biology*, Volume 83
ISSN 0065-2881
https://doi.org/10.1016/bs.amb.2019.08.004

things, sharks have been regarded as: venerated beasts, or gods, of supernatural power (e.g. Beckwith, 1917; Bringhurst, 2011); indiscriminant killing machines (e.g. Benchley, 1974); nuisances impeding fisheries for more profitable species (e.g. COSEWIC, 2011); sources of food, tools, and medicinal products (e.g. Dewhirst, 1980; Newman, 2016); sought after charismatic animals for aquarium displays and ecotourism opportunities (e.g. Grassmann et al., 2017; Mieras et al., 2017); and integral components of marine ecosystems that warrant protection (e.g. Dulvy et al., 2014; Ward-Paige, 2017).

Sharks occupy diverse habitats and ecological niches throughout every ocean on Earth, and some even occur in freshwater systems, and there are over 500 species currently recognized by science (Compagno, 1984a,b, 2001; Dulvy et al., 2014). Their ubiquitous distribution and inherent allure has resulted in countless depictions of sharks in popular culture and over the past several decades scientific inquiry into the nature of these animals has greatly expanded (Heupel et al., 2019; Holland et al., 2019; Huveneers et al., 2018). Recognizing that intense localized fishing pressure and the typical shark life history characterized by long lives, slow maturation, and low fecundity make most shark species vulnerable to exploitation (Musick, 1999; Dulvy et al., 2014). Complicating effective shark management and conservation are complex national and international regulatory and management systems in many jurisdictions that often work in isolation. In many cases, these systems fall short of their conservation goals. All of these factors compound and complicate shark conservation, and recent evaluations suggest that particularly nearshore and large-bodied shark species around the world that most often come into contact with people are at substantially elevated risk of extinction (Dulvy et al., 2014).

In 2004, as a response to concerns about shark fishery exploitation in Puget Sound, local, biennial shark conservation meetings began to be held at the Seattle Aquarium (http://www.seattleaquarium.org/shark-workshop). Initially these conservation workshops focused only on cowsharks (family Hexanchidae), specifically Bluntnose Sixgill (*Hexanchus griseus*) and Broadnose Sevengill (*Notorynchus cepedianus*) sharks. They were also limited in geographic scope, drawing participants only from the west coast of the contiguous United States and Canada. The Cowshark Conservation Workshops progressively grew in breadth and scope to include all species of sharks occurring in the Northeast Pacific Ocean (NEP) off North America, with a concomitant increase in the number and diversity of participants. In 2014 the biennial meeting evolved to become the Northeast Pacific Shark Symposium and continues to convene every other year to discuss advancements in Northeast Pacific shark biology, ecology and conservation (NEPSS; http://www.seattleaquarium.org/shark-workshop).

Recognizing that many of the topics and new knowledge presented at these symposia had not yet been integrated into the published literature, a book was conceived of to share recent findings with a broader audience. Participants in the workshop with expertise in fields ranging from biodiversity and taxonomy, to population genetics, to population modelling and fisheries policy, to shark-focused ecotourism agreed to author chapters resulting in the publication of two volumes in the Advances in Marine Biology (AMB) series in 2017 (Larson and Lowry, 2017a,b). The first of these volumes dealt with systematics and aspects of organismal biology and growth (AMB vol 77), and the second focused on fisheries, captive husbandry, and other population-level elements of human interactions with sharks in the NEP (AMB vol 78). Though the geographic scope of these volumes was initially intended to cover waters from Alaska to Mexico, inclusion of data from Mexico was limited, in part due to the geographic expertise of the contributing authors. These volumes now serve as references for ecologists, natural resource managers, aquarists, and other shark enthusiasts, and have found their way into classrooms and libraries throughout the region.

In 2014 with the expanded scope of the biennial NEPSS workshops researchers and resource managers from Mexico began to attend. In 2016, when the workshop was held at the University of Southern California (USC) Wrigley Marine Science Center on Catalina Island, participation of these scientists increased markedly and it became apparent that there was sufficient interest in, and knowledge about, the sharks of Mexico from these shark biologists south of the border to convene a companion volume about the sharks in the Pacific Ocean off of Mexico. While several substantial treatments of shark fisheries exist for the Pacific waters of Mexico (e.g. Dent and Clarke, 2015) and the scientific literature for the region is increasingly abundant and robust, the NEPSS workshop attendees clearly demonstrated that decades of biological, ecological, and fishery data were available that had not yet been comprehensively reviewed or made broadly available. This volume seeks to address this gap by synthesizing biological and ecological data for, and conservation efforts targeted at, populations of sharks residing in the Pacific Ocean bordering the west coast of Mexico. In our previous volumes (Larson and Lowry, 2017a,b) we defined the NEP as the region from the Bering Strait in the north, down the west coast of North America including the Aleutian Archipelago, the Gulf of Alaska, and the nearshore (inside)

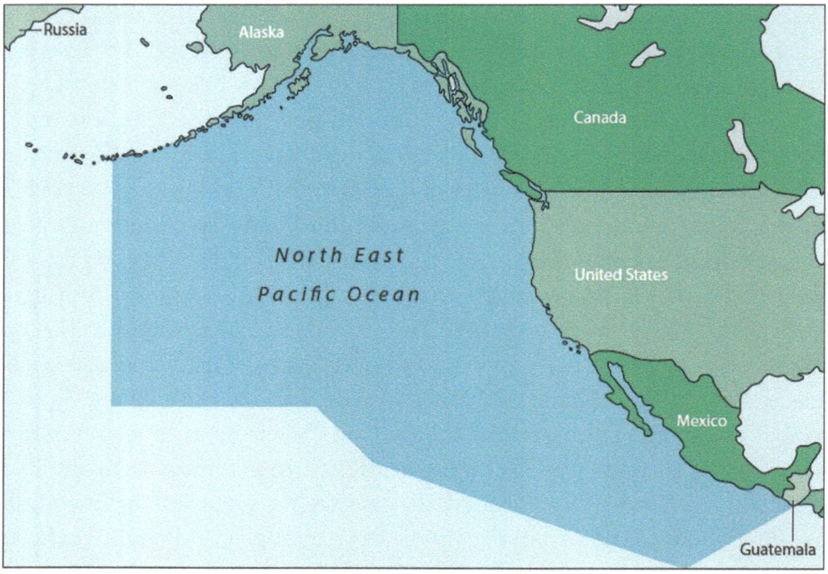


**Fig. 1** Map of the North East Pacific Ocean (NEP) as previously defined in Larson and Lowry (2017a,b).

waters of the Alexander Archipelago; the Straits of Juan de Fuca and Puget Sound; San Francisco Bay, the Southern California Bight; the Gulf of California (or Sea of Cortez); the waters surrounding Baja California; and the waters along the west coast of Mexico to its border with Guatemala (Fig. 1). As noted above, data from Mexico was limited in the first two volumes, however, and for the purposes of this companion volume only the waters in the southern portion of this region are considered, that is, from the U.S.-Mexico Border to approximately the southern border of Guatemala.

## 2. Volumes 83 and 84 content

The chapters of this volume (83) and its companion volume (84) are arranged to provide readers a comprehensive understanding of the current state of the knowledge for sharks occurring in the southern NEP, from the fundamentals of taxonomy through advanced resource management. A dedicated chapter on management and policy is not provided; however, each chapter references diverse regional, national, and international policy documents and actions as germane. Advances in Marine Biology volumes 77 (Larson and Lowry, 2017a) and 78 (Larson and Lowry, 2017b) contained some chapters that are mirrored in this volume and others that are not. In some cases this is because prior chapters incorporated substantial data from Mexican waters, while in other cases it is because comparable research is not currently being conducted in the southern NEP. There are also cases in which the concepts covered are essentially universal and a chapter specific to Mexico would not add appreciably to broader scientific discourse. The goal of volumes 83 and 84 of AMB is to showcase the latest knowledge and research on sharks present in Mexican waters, as well as conservation efforts aimed at these populations, that have not been compiled in a single location to date.

In chapter "Biodiversity and conservation of sharks in Pacific Mexico" Dr Luz Erandi Saldaña-Ruiz of the Center for Scientific Research and Higher Education at Ensenada and her team delve into the biodiversity of sharks known to occur in the southern NEP, and discuss conservation efforts aimed at these species. A thorough understanding of species- and stock-specific variation in life history and distribution is fundamental to successful conservation, and this requires proper identification of these species and their stocks. This chapter complements Ebert et al. (2017) in AMB volume 77, who provided a list of species names and geographic distributions for sharks in the northern NEP, as well as a discussion of the critical role that systematics and taxonomy play in identifying and geographically delineating species, and stocks, for management purposes. Management of sharks differs substantially among jurisdictions within the NEP based on state and country borders, funding, political will, and cultural differences. Some of these differences and complications are highlighted and explained here.

Having identified the species occurring in the region and provided details on their fundamental life history attributes, chapter "Shark ecology, the role of the apex predator and current conservation status" brings into consideration the role of sharks in the ecosystem and how this affects their conservation status. Dr Felipe Galván-Magaña of the National Polytechnic Institute's Interdisciplinary Center of Marine Science (CICIMAR) and his team in La Paz Baja California Sur, Mexico discuss perceptions of shark feeding behaviour and ecology and contrast these with scientific realities, demonstrating how knowledge gaps can lead to misunderstanding of how these complex ecosystems work and how filling these gaps can assist in conservation efforts. This chapter complements Bizzarro et al. (2017) and Reum et al. (2017) in AMB volume 77, which discuss how foraging patterns derived from gut contents and stable isotopes, respectively, can be used to dispel preconceptions about shark feeding and properly place species in their relevant ecological niches. While the level of detail with which we understand the diet of various sharks species varies across the NEP, the positive relationship between knowledge about and the will to conserve a species is demonstrated to hold across international borders.

In recent decades genetic analyses have become increasingly sophisticated with regard to distinguishing species, demographically independent subunits of species, and determining mating patterns such that management and conservation efforts can now be finely tailored. In addition to reviewing how these conservation genetics approaches have been applied to sharks in the southern NEP, in chapter "Conservation genetics of elasmobranchs of the Mexican Pacific Coast, trends and perspectives" Dr Jonathan Sandoval-Castillo of Flinders University and his team also explore how developing sampling techniques, such as collection of environmental DNA (eDNA), might be employed to guide long-term conservation. This chapter complements Larson et al. (2017) in AMB volume 77 and facilitates evaluation of how perceptions of, and support for, genetic research initiatives differ between Mexico and the United States.

While genetic tools can be applied to understand the development of demographic patterns that have formed over generations and to identify unique populations for conservation, understanding short- and long-term movement patterns of individual sharks on seasonal and interannual timeframes is critical to effective conservation and management of existing populations. In the first chapter "Shark movement patterns in the Mexican Pacific: A conservation and management approach" of volume 84 Dr James Ketchum of the University of California, Davis and his team review and synthesize tagging and movement studies conducted over the last several decades for sharks off Pacific Mexico. They demonstrate that while considerable information has been gleaned about where and when sharks can be found, there are still substantial data gaps to be filled before conservation policy can be deemed comprehensive. For some species breeding and pupping grounds remain unknown, and for others mysterious seasonal forays into off shore waters for unquantified reasons have only recently been documented, and have yet to be adequately explained.

Sharks of many species have been exploited by subsistence, recreational, and commercial fisheries in the NEP for centuries or millennia, with uses for the animals ranging widely. Sharks are often taken as bycatch in other fisheries and, in portions of the NEP, discarded at sea (King et al., 2017). In the southern NEP, however, industrialized commercial fishing is not as prevalent as elsewhere in the NEP, with artisanal fisheries playing a larger role in overall catch. In chapter "Fisheries interactions and the challenges of both targeted and non-targeted take in shark conservation" of volume 84 Dr Oscar Sosa-Nishizaki of the Center for Scientific Research and Higher Education at Ensenada and his team review fisheries interactions with sharks in the southern NEP and identify associated conservation challenges. While artisanal fisheries do provide protein-rich food for local communities and is the dominant fishing mode, recent technological and infrastructure developments in the region are causing growing concern for the long-term sustainability of shark exploitation. This chapter complements King et al. (2017), AMB volume 78, who reviewed fisheries in the northern NEP, and draws crucial parallels and distinctions between the two regions that drive broad-scale ocean policy across international boundaries.

Not all uses of sharks are consumptive, and in the warm climes of the southern NEP shark-based ecotourism is prevalent and increasing. From snorkelling and scuba diving to catch-and-release fisheries more people are interacting with sharks face-to-face than ever before in human history. Valuations of resources such as whale sharks have shown that individual sharks can be many times more valuable as ecotourism draws than as food sources. The southern NEP economy has benefited from this ecotourism trend, and in chapter "The economy of shark conservation: The role of ecotourism and citizen science" of volume 84 Dr Andrés M. Cisneros-Montemayor of the University of British Columbia and his team discuss the relative value of sharks for non-consumptive uses. They also discuss how this alternative avenue for education and outreach has been leveraged to create citizen science programmes and engender a life-long conservation mindset in the public at large. This chapter complements Mieras et al. (2017), in AMB volume 78, who discussed diving with and photographing sharks in the northern NEP, among other non-consumptive uses. Regardless of country, the draw of marine megafauna to the thrill seekers and naturalists among us is undeniable.

Bringing the pair of volumes to a close, Dr Oscar Sosa-Nishizaki of the Center for Scientific Research and Higher Education at Ensenada and Dr Felipe Galván-Magaña of the National Polytechnic Institute's CICIMAR in La Paz briefly recap human interactions with sharks in the southern NEP, summarizing the current state of conservation management, and identifying future challenges that may jeopardize the long-term persistence of sharks in the region. While some of these challenges are localized and tractable, such as disorganized or incomplete reporting of shark bycatch in some fisheries, other threats are widespread and of uncertain severity and duration, such as global climate change. This chapter complements and meshes with Lowry (2017), in AMB volume 78, and draws together information from all four volumes to evaluate NEP-wide threats and concerns. It also identifies paths forward, where feasible, to address these issues.

## References

Beckwith, M.W., 1917. Hawaiian shark aumakua. Am. Anthropol. 19 (4), 503–517.

Benchley, P., 1974. Jaws. DoubleDay, New York City. 278 pp.

Bizzarro, J.J., Carlisle, A.B., Smith, W.D., Cortés, E., 2017. Diet composition and trophic ecology of Northeast Pacific Ocean sharks. In: Larson, S.E., Lowry, D. (Eds.), In: Advances in Marine Biology, vol. 77. Academic Press, pp. 111–148.

Bringhurst, R., 2011. A Story as Sharp as a Knife: The Classical Haida Mythtellers and Their World, second ed. D&M Publishers Inc, Vancouver, BC.

Committee on the Status of Endangered Wildlife in Canada (COSEWIC), 2011. Assessment and Status Report on the North Pacific Spiny Dogfish *Squalus suckleyi* in Canada. Committee on the Status of Endangered Wildlife in Canada, Ottawa. 45 pp.

Compagno, L.J.V., 1984a. Sharks of the World: An Annotated and Illustrated Catalogue of Shark Species Known to Date. Part 1. Hexanchiformes to Lamniformes. FAO Fisheries Synopsis 125, 249 pp.

Compagno, L.J.V., 1984b. Sharks of the World. An Annotated and Illustrated Catalogue of Shark Species Known to Date. Part 2. Carcharhiniformes. FAO Fisheries Synopsis 125, 659 pp.

Compagno, L.J.V., 2001. Sharks of the World: An Annotated and Illustrated Catalogue of Shark Species Known to Date. Bullhead, Mackerel and Carpet Sharks (Heterodontiformes, Lamniformes and Orectolobiformes). FAO Species Catalogue for Fishery Purposes 1, vol. 2. FAO, Rome, Italy.

Dent, F., Clarke, S., 2015. State of the Global Market for Shark Products. Food and Agriculture Organization of the United Nations, RomeFAO Fisheries and Aquaculture Technical Paper 590. 187 pp.

Dewhirst, J., 1980. The Yuquot project. In: The Indigenous Archaeology of Yuquot: A Nootkan Outside Village. Parks Canada. History and Archaeology, vol. 39(1). National Historic Parks and Sites Branch, Ottawa, Canada. 365 pp.

Dulvy, N.K., Fowler, S.L., Musick, J.A., Cavanagh, R.D., Kyne, P.M., Harrison, L.R., Carlson, J.K., Davidson, L.N.K., Fordham, D.V., Francis, M.P., Pollock, C.M., Simpfendorfer, C.A., Burgess, G.H., Carpenter, K.E., Compagno, L.J.V., Ebert, D.A., Gibson, C., Heupel, M.R., Livingstone, S.R., Sanciangco, J.C., Stevens, J.D., Valenti, S., White, W.T., 2014. Extinction risk and conservation of the world's sharks and rays. Elife 3, e00590.

Ebert, D.A., Bigman, J.S., Lawson, J.M., 2017. Biodiversity, life history, and conservation of northeastern Pacific chondrichthyans. In: Larson, S.E., Lowry, D. (Eds.), In: Advances in Marine Biology, vol. 77. Academic Press, pp. 9–78.

Grassmann, M., McNeil, B., Wharton, J., 2017. Sharks in captivity: the role of husbandry, breeding, education, and citizen science in shark conservation. In: Larson, S.E., Lowry, D. (Eds.), In: Advances in Marine Biology, vol. 78. Academic Press, pp. 89–119.

Heupel, M.R., Papastamatiou, Y.P., Espinoza, M., Green, M.E., Simpfendorfer, C.A., 2019. Reef shark science—key questions and future directions. Front. Mar. Sci. 6 (12), 1–14.

Holland, K.N., Anderson, J.M., Coffey, D.M., Holmes, B.J., Meyer, C.G., Royer, M.A., 2019. A perspective on future tiger shark research. Front. Mar. Sci. 6 (37), 1–7.

Huveneers, C., Apps, K., Becerril-García, E.E., Bruce, B., Butcher, P.A., Carlisle, A.B., Chapple, T.K., Christiansen, H.M., Cliff, G., Curtis, T.H., Daly-Engel, T.S.,

Dewar, H., Dicken, M.L., Domeier, M.L., Duffy, C.A.J., Ford, R., Francis, M.P., French, G.C.A., Galván-Magaña, F., García-Rodríguez, E., Gennari, E., Graham, B., Hayden, B., Hoyos-Padilla, E.M., Hussey, N.E., Jewell, O.J.D., Jorgensen, S.J., Kock, A.A., Lowe, C.G., Lyons, K., Meyer, L., Oelofse, G., Oñate-González, E.C., Oosthuizen, H., O'Sullivan, J.B., Ramm, K., Skomal, G., Sloan, S., Smale, M.J., Sosa-Nishizaki, O., Sperone, E., Tamburin, E., Towner, A.V., Wcisel, M.A., Weng, K.C., Werry, J.M., 2018. Future research directions on the "elusive" white shark. Front. Mar. Sci. 5 (455), 1–15.

King, J., McFarlane, G.A., Gertseva, V., Gasper, J., Matson, S., Tribuzio, C.A., 2017. Shark interactions with directed and incidental fisheries in the Northeast Pacific Ocean: historic and current encounters, and challenges for shark conservation. In: Larson, S.E., Lowry, D. (Eds.), In: Advances in Marine Biology, vol. 78. Academic Press, pp. 9–44.

Larson, S.E., Daly-Engel, T.S., Phillips, N.M., 2017. Review of current conservation genetic analyses of Northeast Pacific sharks. In: Larson, S.E., Lowry, D. (Eds.), In: Advances in Marine Biology, vol. 77. Academic Press, pp. 79–110.

Larson, S.E., Lowry, D. (Eds.), 2017a. Northeast Pacific shark biology, research and conservation part A, first ed. In: Advances in Marine Biology, vol. 77. Elsevier Ltd., London, p. 230.

Larson, S.E., Lowry, D. (Eds.), 2017b. Northeast Pacific shark biology, research and conservation part A, first ed. In: Advances in Marine Biology, vol. 78. Elsevier Ltd., London, p. 169.

Lowry, D., 2017. Conclusions: the future of shark management and conservation in the Northeast Pacific Ocean. In: Larson, S.E., Lowry, D. (Eds.), In: Advances in Marine Biology, vol. 78. Academic Press, pp. 155–164.

Mieras, P.A., Harvey-Clark, C., Bear, M., Hodgin, G., Hodgin, B., 2017. The economy of shark conservation in the Northeast Pacific: the role of ecotourism and citizen science. In: Larson, S.E., Lowry, D. (Eds.), In: Advances in Marine Biology, vol. 78. Academic Press, pp. 121–153.

Musick, J.A., 1999. Criteria to define extinction risk in marine fishes: the American Fisheries Society initiative. Fisheries 24 (12), 6–14.

Newman, S.E., 2016. Sharks in the jungle: real and imagined sea monsters of the Maya. Antiquity 90 (354), 1522–1536.

Reum, J.C.P., Williams, G.D., Harvey, C.J., 2017. Stable isotope applications for understanding shark ecology in the Northeast Pacific Ocean. In: Larson, S.E., Lowry, D. (Eds.), In: Advances in Marine Biology, vol. 78. Academic Press, pp. 149–178.

Ward-Paige, C.A., 2017. A global overview of shark sanctuary regulations and their impact on shark fisheries. Mar. Policy 82, 87–97.

CHAPTER TWO

# Biodiversity and conservation of sharks in Pacific Mexico

Luz Erandi Saldaña-Ruiz[a,*], Emiliano García-Rodríguez[a], Juan Carlos Pérez-Jiménez[b], Javier Tovar-Ávila[c], Emmanuel Rivera-Téllez[d]

[a]Departamento de Oceanografía Biológica, Centro de Investigación Científica y de Educación Superior de Ensenada (CICESE), Ensenada, Baja California, México
[b]El Colegio de la Frontera Sur (ECOSUR), Lerma, Campeche, México
[c]Instituto Nacional de Pesca (INAPESCA), Centro Regional de Investigacion Pesquera (CRIP), La Cruz de Huanacaxtle, Nayarit, México
[d]Coordination of Mexican CITES Scientific Authority, National Commission for the Knowledge and Use of Biodiversity (CONABIO), Mexico City, México
*Corresponding author: e-mail address: luzerandisr@gmail.com

## Contents



## Abstract

Mexico is home to a broad biodiversity of shark species, and more than half of the sharks in Mexican waters are distributed in the Mexican Pacific, with over 62 species recorded. This high biodiversity is the result of numerous and diverse marine and coastal environments, including the dynamic Mexican seas, where circulation and spatial variation of oceanic currents is complex, and the seasonal variation of isotherms can be substantial. In the Mexican Pacific we can distinguish some patterns of species distribution, with temperate water and subtropical species found in the northern regions, and tropical

*Advances in Marine Biology*, Volume 83
ISSN 0065-2881
https://doi.org/10.1016/bs.amb.2019.08.001

conditions and species found in the south. Due to the blending of cold and warm waters, however, we can find a mixture of subtropical and tropical sharks in northern regions seasonally, off the west coast of the Baja California Peninsula and in the Gulf of California, and these areas contain the highest shark species richness. In this chapter we described the shark species biodiversity occurring in the Mexican Pacific, review their conservation status in a regional and global context, and summarize the main conservation measures and issues associated with their management.

## 1. Introduction

Sharks belong to the taxonomic class Chondrichthyes, also known as cartilaginous fishes. These fishes have flexible cartilage with no bone present in the skeleton, fins, or scales (Ebert et al., 2013; Klimley, 2013), except for trace amounts of acellular bone at the base of the denticles. In the Mexican Pacific (MP), the complex oceanographic features and large variety of ecosystems have led to great fish biodiversity, including sharks (Ehemann et al., 2018; Espinosa-Pérez, 2014; Lara-Lara et al., 2008). Sharks occupy a wide range of habitats, including coastal waters, the open sea, and the deep ocean. However, as in the rest of the northeast Pacific Ocean, most species occur along continental shelves and slopes (Ebert et al., 2017), and around islands, with some endemic species restricted to small areas or in deep waters (Cailliet et al., 2005).

Most species (44.4%) in the MP, including sharks of the genera *Heterodontus*, *Carcharhinus*, *Sphyrna*, and *Mustelus*, occur on continental shelves from the intertidal zone, such as bays and estuaries, along coastal nearshore areas up to 200 m deep (Castro, 2011a; Del Moral-Flores et al., 2015a; Ebert et al., 2013, 2017). Almost 30% of the MP sharks inhabit deep waters of the continental slope, the external platform, and the abyssal region, especially on the west coast of the Baja California and in the Gulf of California (Castro-Aguirre and Balart, 1996; Mejía-Mercado et al., 2013). There are pelagic sharks, oceanic and semi-oceanic, of the families Rhincodontidae, Alopiidae, Cetorhinidae, Lamnidae, Megachasmidae, Carcharhinidae, and Sphyrnidae. Some pelagic sharks are rare, such as Bigeye Sand Tiger, *Odontaspis noronhai*, and Crocodile Shark, *Pseudocarcharias kamoharai*, and others are abundant, such Blue shark, *Prionace glauca*, and Silky shark, *Carcharhinus falciformis* (Compagno, 2008; Del Moral-Flores et al., 2015a,b).

In this chapter, we describe the general patterns of diversity and distribution of sharks in the MP through the synthesis and updating of published literature. We also review the conservation status of the species and policies related to the conservation and management of shark species. We discuss transboundary shark conservation, as well as the role of nursery areas and protected areas in shark conservation in the MP.

## 2. The Mexican Pacific ecosystem

The MP extends from approximately 32° 31′ N, 117° 07′ W to about 14° 32′ N 92° 14′ W, from the north of the Baja California peninsula to the Gulf of Tehuantepec (Fig. 1). The MP has a variety of physical characteristics and diverse marine ecosystems, and it has a highly variable bathymetry with depths of more than 2000 m in most of the exclusive economic zone and a maximum depth recorded off the coast of Chiapas and Oaxaca in the Tehuantepec pit of more than 6000 m (Espinosa, 2004; Lara-Lara et al., 2008). The MP encompasses four large marine ecosystems based on physical and biological characteristics: the west coast of the

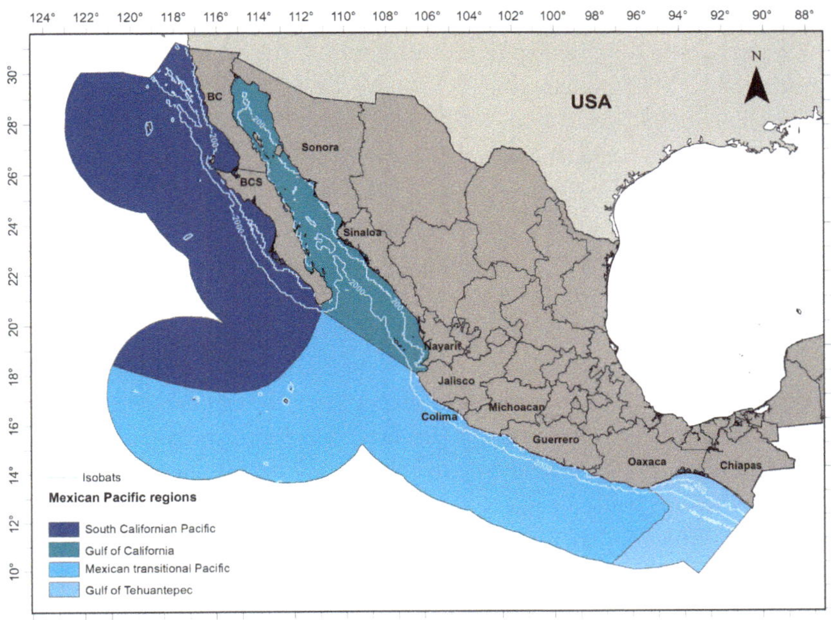


**Fig. 1** Delimitation of the Mexican Pacific, its regions and isobaths. *Data from CEC, 2009. Marine Ecoregions of North America, 2008, Vector Digital Data. Canada CEC, Montreal, Quebec, Canada. http://www.cec.org/tools-and-resources/map-files/marine-ecoregions-level-ii (Accessed 01 Jan 2019), and INEGI, Lugo-Hupb J., Vidal-Zepeda, R., Fernández-Equiarte, A., Gallegos-García, A., Zavala H, J.Y Otros. (1990). Hipsometría y Batimetría, I.1.1. Atlas Nacional de México. vol. I. Escala 1:4000000. Instituto de Geografía, UNAM, México.*

Baja California Peninsula where tropical and temperate conditions intermingle, known as the South Californian Pacific (SCP) the Gulf of California (GOC) between the peninsula and the mainland; the Mexican transitional Pacific (MTP) where faunal composition varies seasonally; and the Gulf of Tehuantepec (GOT), which is tropical (Fig. 1) (Wilkinson et al., 2009).

The distribution of most shark species is predominantly influenced by temperature, limiting many geographic ranges (Carrier et al., 2010). Shark diversity in the MP is higher in the north, where a mixture of cold and warm temperatures occurs (Durazo et al., 2007). In the SCP and GC regions, more than 80% of the species distributed in the MP can be found. In the SCP there are eight orders, 22 families, 35 genera, and 53 species, while in the GOC there are eight orders, 17 families, 30 genera, and 52 species. In the southern regions, the MTP and GOT, temperate species are missing and only 50% of the species are present: five orders, 11 families, 17 genera, and 31 species (Del Moral-Flores et al., 2015a; Ebert et al., 2013; Ehemann et al., 2018).

The SCP is the southern limit of the California Current System, a transition zone with high biodiversity due to the mixture of temperate and nutrient-rich cold waters from the California Current and warm waters from the south (Castro et al., 2017, Durazo et al., 2007). In the SCP are found shark species of subpolar and cold-temperate waters like Salmon Shark, *Lamna ditropis*, and Basking Shark, *Cetorhinus maximus*, as well as warm-temperate or even subtropical species like Horn Shark, *Heterodontus francisci*, Narrowtooth Shark, *Carcharhinus brachyurus*, Dusky Shark, *Carcharhinus obscurus*, Smalltail Shark, *Carcharhinus cerdale*, and Smooth Hammerhead, *Sphyrna zygaena* (Castro, 2011b; De La Cruz-Agüero et al., 1997; Del Moral-Flores et al., 2015a; Ebert et al., 2013; García-Rodríguez, 2012; Reyes-Bonilla et al., 2010; Santos del Prado and Peters, 2005; Stevenson et al., 2007). In this region are pelagic oceanic and semi-oceanic sharks from the genera *Prionace*, *Isurus*, *Pseudocarcharias*, and *Sphyrna*, which also are widespread species with circumglobally distributions; and coastal species like Pacific Angel Shark, *Squatina californica*, Brown Smoothhound, *Mustelus henlei*, and Grey Smoothhound, *Mustelus californicus* (Cartamil et al., 2011; Castillo-Géniz et al., 2016; Del Moral-Flores et al., 2015a; Ebert et al., 2013; García-Rodríguez, 2012; Long and Seigel, 1997; Reyes-Bonilla et al., 2010). Rare deepwater species of the families Pentanchidae Chlamydoselachidae, Hexanchidae, Dalatiidae, Etmopteridae, and Somniosidae have also been registered in this region (Castro-Aguirre et al., 2003; Espinosa-Pérez et al., 2004; Reyes-Bonilla et al., 2010; Ruiz-Campos et al., 2010; Stevenson et al., 2007). Within this region, an aggregation site of White Shark, *Carcharodon carcharias* has been reported at Guadalupe Island, where white sharks spend approximately half of the year and then migrate to a large, offshore pelagic environment (Domeier and Nasby-Lucas, 2008).

The GOC region is a semi-enclosed sea in the eastern North Pacific Ocean that is 1130 km long and approximately 200,000 $km^2$ in extent, with depths greater than 3000 m (Lluch-Cota et al., 2007). This region is located between the Baja California Peninsula and the States of Sonora, Sinaloa, and Nayarit, and its southern limit is defined as an imaginary line connecting Cabo San Lucas (Baha California Sur [BCS]) and Cabo Corrientes (Jalisco) (Álvarez-Borrego, 2010; Brusca et al., 2005) (Fig. 1). This region has variable high sea surface temperature and high primary productivity (mostly in the north and central zones) during winter and spring, due to oceanographic processes like gyres, fronts, and upwellings (Álvarez-Romero et al., 2013). These dynamics processes contribute to a high diversity of sharks and another ichthyofauna in the GOC (Lara-Lara et al., 2008; Del Moral-Flores et al., 2016a).

In the northern portion of the GOC, coastal habitat species predominate, from subtropical sharks like Brown Smoothhound, *Mustelus henlei*, and Leopard Shark, *Triakis semifasciata*, to tropical sharks like Mexican Horn Shark, *Heterodontus mexicanus*, Pacific Angel Shark, *Squatina californica*, and Pacific Sharpnose Shark, *Rhizoprionodon longurio* (De Cruz-Agüero et al., 1996; Ebert et al., 2013; Smith et al., 2009). However, pelagic oceanic, and semi-oceanic species of the genera *Alopias*, *Carcharhinus*, and *Isurus* are also distributed in the northern GOC (Bizzarro et al., 2009b; Smith et al., 2009). Shark diversity is higher in the southern GOC, due to the great variety of habitats related to high water temperatures and connectivity with Pacific waters (Álvarez-Borrego, 2010; Brusca et al., 2005). Sharks of coastal and oceanic habitats, and subtropical and tropical distribution, are found in the southern GOC, such as species of the *Prionace*, *Rhizoprionodon*, *Sphyrna*, *Nasolamia*, *Negaprion*, and *Alopias* genera (Bizzarro et al., 2009a; Corro-Espinosa 2016; Márquez-Farías, 2000; Mondragón-Sánchez, 2011; Pérez-Jiménez et al., 2005). The southern GOC is an important aggregation site for several species. At submarine mountains at Isla Espiritu Santo and Banco Gordo, near to Bahía de La Paz in Baja California Sur, aggregations of Scalloped Hammerhead, *Sphyrna lewini*, and Whale Shark, *Rhincodon typus*, have been reported, and this site is considered a hotspot of biodiversity (González-Rodríguez et al., 2018; Klimley, 1981; Ramírez-Macias et al., 2012). The Whale Shark has also been reported along coastal areas from the GOC, at Bahia de la Paz (Baja California Sur), Bahía de Los Angeles,

San Luis Gonzaga (Baja California), and the central coast of Nayarit (Ramírez-Macias et al., 2012). Deepwater species like Largenose Catshark, *Apristurus nasutus*, Peppered Catshark, *Galeus piperatus*, Filetail Catshark, *Parmaturus xaniurus*, Bluntnose Sixgill Shark, *Hexanchus griseus*, and Prickly Shark, *Echinorhinus cookei*, are also present in the GOC region (Acevedo-Cervantes et al., 2009; Bizzarro et al., 2009a,b; Hubbs and Taylor, 1970; Mejía-Mercado et al., 2013). Even in extremely low-oxygen marine habitats of the deep waters of the GOC, species like Lollipop Catshark, *Cephalurus cephalus*, thrive (Aguirre-Villaseñor and Salas-Singh, 2012; Balart et al., 2000; Gallo et al., 2019; Graeber, 1974).

In the GOC, around 10% of fishes with a variety of phylogenetic and geographic affinities are endemic, and among chondrichthyan species approximately 4.2% are endemic (Del Moral-Flores et al., 2016a; Hastings et al., 2010). The endemic shark species in the GOC are Peppered Catshark, *Galeus piperatus,* and Whitemargin Smoothhound, *Mustelus albipinnis*. However, research on these deepwater sharks is scarce, and Whitemargin Smoothhound is probably mistaken for other *Mustelus* species present in the region (Ebert et al., 2013; Palacios-Salgado et al., 2012). For all the above, is very likely that these species could have a wider distribution, decreasing the endemism of the region.

The MTP is a productive region influenced by the southern end of the California Current and the surface current from the Gulf of California during winter and by the North Equatorial Counter current in the summer (Badan, 1997; De la Lanza, 1991; Filonov et al., 2000). Due to these seasonal changes in oceanographic conditions, a seasonally variable combination of subtropical and tropical shark species can be found. There is a clear predominance of tropical species like Galapagos Shark, *Carcharhinus galapagensis*, Bull Shark, *Carcharhinus leucas*, Silvertip Shark, *Carcharhinus albimarginatus*, Pacific Smalltail Shark, *Carcharhinus cerdale*, Scalloped Bonnethead, *Sphyrna corona*, Bonnethead, *Sphyrna tiburo*, Great Hammerhead, *Sphyrna mokarran*, Mexican Horn Shark, *Heterodontus mexicanus*, Sicklefin Smoothhound, *Mustelus lunulatus*, Whitenose Shark, *Nasolamia velox*, Lemon Shark, *Negaprion brevirostris*, and Pacific Sharpnose Shark, *Rhizoprionodon longurio* (Castro, 2011b; Cruz et al., 2011; Del Moral-Flores et al., 2016b; Santana-Hernández et al., 2013; Vélez-Marín and Márquez-Farías, 2009). There are also shark species whose preferences are for temperate waters, such as members of the genera *Alopias*, *Sphyrna*, *Carcharhinus*, and *Echinorhinus* (Santana-Hernández and Flores-Valdez, 2014; Santana-Hernández et al., 2014). Furthermore, in this region, there are pelagic sharks of worldwide distribution, like Silky Shark, Blue Shark, and Shortfin Mako (Cruz et al., 2011; Santana-Hernández and Flores-Valdez, 2014; Vélez-Marín and Márquez-Farías, 2009). In the Revillagigedo Islands small aggregations of Whale Shark, and aggregations of Whitetip Reef Shark, *Triaenodon obesus* and Galapagos Shark, *Carcharhinus galapagensis* have been observed associated with coral reefs; the aggregation behaviour of these reef-associated species has been recorded before in the Northwestern Hawaiian Islands (Cortés-Fuentes, 2018; Holzwarth et al., 2006; Ramírez-Macias et al., 2012).

The GOT is located in waters of the Eastern Tropical Pacific current; it has high seasonal variability in productivity due to variations in upwelling and nutrient inputs from rivers present in coastal areas of Chiapas (Gallegos-García and Barberán-Falcón, 1998; Trasviña et al., 1995, 1999). The GOT is a tropical region since it does not receive the influence of the California Current; hence, many tropical species of the genera *Nasolamia*, *Carcharhinus*, *Mustelus*, *Galeocerdo*, *Rhizoprionodon*, *Ginglymostoma*, and *Sphyrna* are distributed in this region. In the GOT, species of a wide distribution of the genera *Carcharhinus*, *Alopias*, *Prionace*, and *Isurus* are also found (Alejo-Plata et al., 2007; Castillo Géniz et al., 2002; Morales-Pacheco et al., 2006; Soriano-Velásquez et al., 2006). Small aggregations of Whale Sharks off the coasts of Salina Cruz, Oaxaca have also been observed (Ramírez-Macias et al., 2012).

## 3. Sharks biodiversity in the Mexican Pacific

The Mexican Pacific has a diversity of 62 shark species, approximately 55% of the total shark species in Mexico (of a total of 111 species) (Table 1).

**Table 1** Known biodiversity of shark species within Mexico and Pacific Mexico by order, family, genus, and species.

| Region | Order | Family | Genus | Species |
|---|---|---|---|---|
| Mexico | 8 | 23 | 48 | 111 |
| Pacific Mexico | 8 | 22 | 37 | 63 |
| South Californian Pacific | 8 | 22 | 34 | 53 |
| Gulf of California | 8 | 22 | 30 | 53 |
| Mexican transitional Pacific | 5 | 12 | 17 | 30 |
| Gulf of Tehuantepec | 5 | 12 | 17 | 31 |

This is 13% of the valid known shark species in the world (Ebert et al., 2013; Ehemann et al., 2018; Weigmann, 2016). In the MP there are eight orders, 22 families, 37 genera, and 62 species (Castro, 2011a; Del Moral-Flores et al., 2015a,b; Ebert et al., 2013; Ehemann et al., 2018). The Mexican Pacific shark diversity is higher than other regions of the world such as in the Mediterranean (49 species), and the Northeast Pacific (43 species), including the Pacific coast of Canada and the United States (Ebert et al., 2017). A checklist of current shark species in the MP region is provided in Appendix. The eight orders distributed in the MP are described in the following paragraphs; shark species geographic ranges are from Ebert et al. (2013), Castro (2011a), and Del Moral-Flores et al. (2015a,b); and the systematic arrangement is based on Nelson et al. (2016).

*Heterodontiformes* (*Horn sharks*). This order comprises just one family, Heterodontidae, a single genera *Heterodontus*, and nine species, of which only two occur in the MP. The Horn Shark, *Heterodontus francisci*, is distributed in the SCP and GOC, while the Mexican Horn Shark, *H. mexicanus*, tolerates temperate-tropical waters and its distribution includes the south of the SCP, GOC, MTP, and GOT (Fig. 2) (Del Moral-Flores et al., 2015a, 2016b).

*Orectolobiformes* (*Carpetsharks*). This group has seven families and 44 species; in the MP only two species from the Ginglymostomatidae and Rhincodontidae families are found (Del Moral-Flores et al., 2015a,b; Weigmann, 2016). The Pacific Nurse Shark, *Ginglymostoma unami*, previously considered as *G. cirratum*, a nurse shark now restricted to the Atlantic, is distributed throughout the MP (Castro, 2011a; Del Moral-Flores et al., 2015b; Ebert et al., 2013) (Fig. 2). The Whale Shark is a highly migratory filter-feeding species (Ebert et al., 2013). In the MP, Whale Shark aggregation sites are reported in the GOC, in Bahia de Los Angeles, Bahia de La Paz, and Gorda Banks, and Isla Espiritu Santo; and small aggregations at San Blass, Nayarit, Salina Cruz, Oaxaca, and Revillagigedo Islands (Ramírez-Macias et al., 2012).

*Lamniformes* (*Mackerel sharks*). This group comprises seven families and 15 species, of which six families and 12 species occur in the MP (Ebert et al., 2013; Weigmann, 2016). The family Alopiidae includes three species (Ebert et al., 2013). In the MP, the Pelagic Thresher, *Alopias pelagicus*, is frequently reported (Castro, 2011a). Reports of Basking Shark, *Cetorhinus maximus*, the only species of the family Cetorhinidae, in the MP are very scarce, with a single confirmed record from the SCP (Castro, 2011a; Sandoval-Castillo et al., 2008). Of the five species of the family Lamnidae,

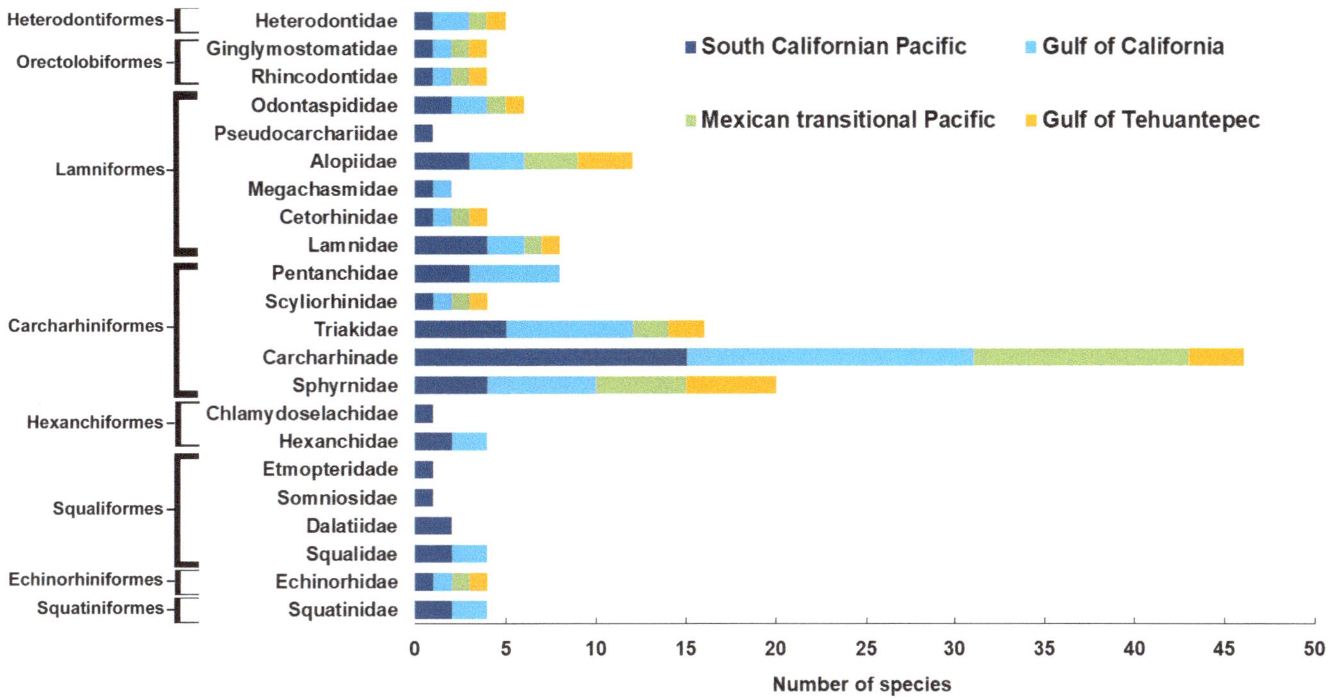


**Fig. 2** The number of shark species by family within each region of the Mexican Pacific. The taxonomic arrangement is based on Nelson et al. (2016).

four are distributed in the MP: White Shark, *Carcharodon carcharias*, with more frequent reports in the SCP and GOC regions (Álvarez-León and Castro-Aguirre, 1983; Galván-Magaña et al., 2010; Kato, 1965; Oñate-González et al., 2015; Weng et al., 2007, 2012); Shortfin Mako, *Isurus oxyrinchus*, mostly observed in the SCP and the GOC (Bizzarro et al., 2009a,b; Ramírez-Amaro et al., 2013; Smith et al., 2009); and Longfin Mako, *I. paucus*, and Salmon Shark, *Lamna ditropis*, with records only in the SCP region (Castro, 2011a; Del Moral-Flores et al., 2015a; Ebert et al., 2013; Ruiz-Campos et al., 2010). The Megamouth Shark, *Megachasma pelagios*, Smalltooth Sand Tiger Shark, *Odontaspis ferox*, Bigeye Sand Tiger Shark, *O. noronhai*, and Crocodile Shark, *Pseudocarcharias kamoharai*, from the Megachasmidae, Odontaspididae, and Pseudocarchariidae families, respectively, are also reported in the SCP region of the MP, with very few records (Fig. 2) (Castillo-Géniz et al., 2012; Castro, 2011a; Ebert et al., 2013).

*Carcharhiniformes* (*Ground sharks*). This order is the largest and most diverse group with eight families and at least 291 species (Weigmann, 2016). In the MP, five families are documented to occur. The Carcharhinidae family comprises many common coastal species like Galapagos Shark, *Carcharhinus galapagensis*, Bull Shark, *C. leucas*, Dusky Shark, *C. obscurus*, Whitenose Shark, *Nasolamia velox*, and Pacific Sharpnose Shark, *Rhizoprionodon longurio*, and oceanic pelagic species like Silky Shark, *C. falciformis*, Blue Shark, *Prionace glauca*, and Oceanic Whitetip Shark, *C. longimanus* (Del Moral-Flores et al., 2015a; Ebert et al., 2013). The species from the Pentanchidae family are deep-water sharks (in more than 900 m depth) like the Brown Catshark, *Apristurus brunneus*, only reported in the SCP region; Longnose Catshark, *A. kampae*, and Filetail Cat Shark, *Parmaturus xaniurus* in the SCP and GOC; and Largenose Catshark, *A. nasutus*, Lollipop Catshark, *Cephalurus cephalus*, and Peppered Catshark, *Galeus piperatus*, reported only in the GOC region (Balart et al., 2000; Castro-Aguirre, 1981; Ebert et al., 2013; Hubbs and Taylor, 1970; Mejía-Mercado et al., 2013; Weigmann, 2016). Only one species of the family Scyliorhinidae, the Swell Shark, *Cephaloscyllium ventriosum*, is documented in the MP (Ebert et al., 2013; Weigmann, 2016). The Sphyrnidae family, hammerhead sharks, comprises eight species, of which six are distributed in the MP, the most common being the Scalloped Hammerhead, *Sphyrna lewini*, and Smooth Hammerhead, *S. zygaena*. The family Triakidae is represented by the genus *Mustelus*, *Triakis*, and *Galeorhinus* (Fig. 2) (Bizzarro et al., 2009a,b,c; Del Moral-Flores et al., 2015a; Ebert et al., 2013; Weigmann, 2016).

*Hexanchiformes (Frilled sharks and cow sharks)*. This order comprises two families and six species. In the MP, the species *Chlamydoselachus anguineus* (family Chlamydoselachidae) is reported in the SCP, and the Bluntnose Sixgill Shark, *Hexanchus griseus* and Broadnose Sevengill Shark, *Notorynchus cepedianus* (family Hexanchidae) are recorded in the SCP and the GOC (Bizzarro et al., 2009a,b; Castro-Aguirre et al., 2003; Ebert et al., 2013; Ruiz-Campos et al., 2010).

*Echinorhiniformes (Bramble sharks)*. This order consists of only two species, both in family Echinorhinidae. This order is characterized by large skin denticles, a stout cylindrical body, five gills slits in front of the pectoral fins, two similar-sized dorsal fins (without spines) close to the caudal fin, and the absence of an anal fin (Ebert et al., 2013). In the MP, only the Prickly Shark, *Echinorhinus cookei*, is distributed in the SCP and GOC regions (Fig. 2) (Long et al., 2011; Ramírez-Amaro et al., 2013; Ruiz-Campos et al., 2010).

*Squaliformes (Dogfish sharks)*. This order includes six families and 130 species of small to medium size sharks (Ebert et al., 2013; Weigmann, 2016), of which only four families and five species are distributed in MP, in the SCP region: Pygmy Shark, *Euprotomicrus bispinatus* and Cookiecutter Shark, *Isistius brasiliensis* (family Dalatiidae); Combtooth Dogfish, *Centroscyllium nigrum* (family Etmopteridae); Pacific Sleeper Shark, *Somniosus pacificus* (family Somniosidae); and Pacific Spiny Dogfish, *Squalus suckleyi* (family Squalidae). The Pacific Spiny Dogfish is also reported in the GOC region (Fig. 2) (Castro-Aguirre and García-Domínguez, 1988; Del Moral-Flores et al., 2015a; Ebert et al., 2010; Espinosa-Pérez et al., 2004; Guzmán and Meráz-Munguía, 2013; Ramírez-Amaro et al., 2013; Reyes-Bonilla et al., 2010; Ruiz-Campos et al., 2010).

*Squatiniformes (Angel sharks)*. The species in this order are medium-sized with a flattened body (Ebert et al., 2013). This order has only one family, Squatinidae, and one genus, *Squatina* containing 20 species (Weigmann, 2016). In the MP, only one species is reported, the Pacific Angel Shark, *Squatina californica*. This benthic shark is very common in the SCP and GOC regions (Fig. 2) (Abitia-Cárdenas et al., 1994; Applegate et al., 1979; Buckhorn, 2012; Galvan-Magaña et al., 1996; Ramírez-Amaro et al., 2013; Rodríguez-Romero et al., 2008).

## 4. Conservation

### 4.1 Conservation status of sharks from the Mexican Pacific

The science-based IUCN Species Survival Commission (SSC) systematically evaluates the relative extinction risk for exploited marine fishes, including sharks, skates, rays, and chimaeras (Chondrichthyes), using the Red List Categories and Criteria of the International Union for the Conservation of Nature (IUCN) (Camhi et al., 2009; Dulvy et al., 2014). According to an evaluation by the IUCN Shark Specialist Group, around 20% of the rays and sharks worldwide are threatened (in the categories of Vulnerable, Endangered, or Critically Endangered) (IUCN-SSG, 2019). Of the 62 shark species in the Mexican Pacific, 31% are threatened. Most species, globally (41%) and for Mexico (30%), have a Data Deficient status due to the lack of information to perform an assessment (Table 2).

Within the MP, the majority of threatened shark species are pelagic with semi-oceanic distribution. Many of these species have a high overlap with fisheries in the MP like Scalloped Hammerhead, Silky Shark, Pelagic Thresher, Smooth Hammerhead, and Dusky Shark (Saldaña-Ruiz et al., 2017; Vélez-Marín and Márquez-Farías, 2009). Three species are listed as Endangered (the Whale Shark, Scalloped Hammerhead, and Great Hammerhead), and 14 as Vulnerable (Table 2). Most of the deep-water sharks in the MP are listed as Data Deficient (Odontaspididae, Pentanchidae, Hexanchidae, Etmopteridae, and Somniosidae families), as well as coastal shark species like Horn Shark, Mexican Horn Shark, and Pacific Sharpnose Shark, common in artisanal fisheries in the MP (Morales-Pacheco et al., 2006; Pérez-Jiménez et al., 2005; Saldaña-Ruiz et al., 2017). Deepwater shark species are often taken as bycatch in deep-sea fishing operations in the MP (Cartamil et al., 2011). Nevertheless, their interaction with fisheries and the status of their populations is unknown. Finally, there are two species, Pacific Nurse Shark and Pacific Smalltail Shark that have not been assessed by the IUCN, which can be attributed to recent taxonomic readjustments (Del Moral-Flores et al., 2015b; Ehemann et al., 2018).

In the MP, 3 sharks listed as Endangered and 11 species listed as Vulnerable are also in Appendix II of the Convention on International Trade in Endangered Species of Wild Fauna and Flora (CITES) (Table 3). CITES has the purpose of regulating international trade of commercially exploited species and complements the national and regional management of species of signatory nations (Parties) (Vincent et al., 2014). This regulation requires a confirmation that the trade is not detrimental for the survival of the species (a Non-Detriment Finding, NDF), before the issuance of CITES export permits for these species, the NDF is formulated by the CITES Scientific Authority of the exporting Party, which stands for the sustainability of their harvest. Species threatened by international trade are proposed for their

**Table 2** Number and percent of Mexican Pacific sharks in IUCN Red List categories by their main habitats.

| Habitat | Species % | Threatened (%) | CR (%) | EN (%) | VU (%) | NT (%) | LC (%) | DD (%) | NE (%) |
|---|---|---|---|---|---|---|---|---|---|
| Coastal | 27 (43.5) | 3 (4.8) | 0 | 0 | 3 (4.8) | 8 (12.9) | 7 (11.3) | 8 (12.9) | 2 (3.2) |
| Pelagic oceanic | 6 (9.6) | 5 (8.1) | 0 | 0 | 5 (8.1) | 2 (3.2) | | 0 | 0 |
| Pelagic semi-oceanic | 11 (17.7) | 7 (11.3) | 0 | 3 (4.8) | 4 (6.5) | 0 | 2 (3.2) | 0 | |
| Deepwater | 18 (29.0) | 2 (3.2) | 0 | 0 | 2 (3.2) | 1 (1.6) | 6 (9.7) | 9 (14.5) | 0 |
| Total | 62 | 17 (27.4) | 0 | 3 (4.8) | 14 (22.6) | 11 (17.7) | 15 (24.2) | 17 (27.4) | 2 (3.2) |

CR, Critically Endangered; EN, Endangered; VU, Vulnerable; NT, Near Threatened; LC, Least Concern; DD, Data Deficient. The number of species considered to be Threatened is the sum of Critically Endangered (CR), Endangered (EN), and Vulnerable (VU) species. All IUCN Red List assessments are available online: www.iucnredlist.org.

**Table 3** Mexican Pacific shark species listed in CITES, the Appendix where listed, the year and the Conference of the Parties where listing was accepted, and the date of entry into force are presented (https://www.cites.org/eng/prog/shark/history.php).

| Scientific name | CITES Appendix | Conference of parties (year listed) | Entry into force |
|---|---|---|---|
| *Rhincodon typus* | II | CoP12 (Chile, 2002) | February 13, 2003 |
| *Alopias pelagicus* | II | CoP17 (South Africa, 2016) | October 4, 2017 |
| *Alopias superciliosus* | II | CoP17 (South Africa, 2016) | October 4, 2017 |
| *Alopias vulpinus* | II | CoP17 (South Africa, 2016) | October 4, 2017 |
| *Cetorhinus maximus* | II | CoP12 (Chile, 2002) | February 13, 2003 |
| *Carcharodon carcharias* | II | CoP13 (Thailand, 2004) | January 12, 2005 |
| *Carcharhinus falciformis* | II | CoP17 (South Africa, 2016) | October 4, 2017 |
| *Carcharhinus longimanus* | II | CoP16 (Thailand, 2013) | September 14, 2014 |
| *Sphyrna lewini* | II | CoP16 (Thailand, 2013) | September 14, 2014 |
| *Sphyrna mokarran* | II | CoP16 (Thailand, 2013) | September 14, 2014 |
| *Sphyrna zygaena* | II | CoP16 (Thailand, 2013) | September 14, 2014 |

inclusion in Appendices I, II, or III of CITES during the Conference of the Parties (CoP), the listing of species requires a supporting vote the majority of parties present at the CoP (Vincent et al., 2014).

## 4.2 Management and conservation measures

### *4.2.1 Plans of action*

Several management and conservation measures have been developed in Mexico in the last decades to promote the sustainability of sharks. Most of those measures were developed based on the National Plan of Action for the Management and Conservation of Sharks, Rays, and Related Species in Mexico (NPOA-Mexico) (CONAPESCA-INP, 2004), in accordance with the International Plan of Action for the Conservation and Management of Sharks (IPOA-Sharks) of the Food and Agriculture Organization (FAO) of the United Nations (FAO, 1999).

Base on the identification of main threats to shark populations, the NPOA-Mexico states that shark catches should be sustainable, establishing as a priority the identification and protection of their critical habitats, particularly for vulnerable or threatened species or stocks. The plan encourages development of measures to minimize bycatch and discards, as well as the full utilization of dead sharks. It also promotes the compilation of biological information on sharks, as well as improving the collection of catch and landings data, facilitating the comprehensive reporting of biological and trade data (CONAPESCA-INP, 2004).

In 2015, the governments of Canada, Mexico, and the United States initiated a collaborative project through the Commission for Environmental Cooperation (CEC) to strengthen the conservation and sustainable trade of North American taxa that are included in Appendix II of the CITES. In 2017, a stakeholder consultation was held in Mexico City to develop a plan of action to promote sustainable trade and conservation of the priority shark species. This action plan includes information on the distribution, conservation status, trade, and commercial pricing of eight priority sharks, as well as the identification challenges for CITES enforcement. The plan serves as a guide for reducing illegal and unsustainable harvest and trade; improving biological knowledge to allow science-based management decisions; and promoting traceability, species conservation, and livelihoods of stakeholders, throughout a total of 17 recommended collaborative actions (CEC, 2017).

### 4.2.2 National fisheries and environmental laws, and management actions

The General Law for Sustainable Fishing and Aquaculture (LGPAS) regulate marine fisheries in Mexico (including those for sharks), and was enacted in 2007 (DOF, 2007a). The LGPAS is implemented by the National Commission of Fishing and Aquaculture (CONAPESCA). Similarly, the Wild Life General Law (LGVS) regulates the sustainable use, conservation, and management of native wild animals and plants (terrestrial and aquatic) that are at risk (DOF, 2014a). The LGVS establishes the national policy for wildlife protection and sustainable use, via the Official Mexican Standard NOM-059-SEMARNAT-2010 (NOM-059).

NOM-059 is the reference instrument of the LGVS, which defines the criteria that must be met for a species to be considered at risk, establishing four categories: Probably Extinct (in the wild), Endangered, Threatened, and Subject to Special Protection (DOF, 2010a). Three shark species are included in NOM-059 and are listed as 'Threatened': White Shark, Whale Shark, and Basking Shark, prohibiting their fishing in Mexican waters. Furthermore, in February 2014, fisheries and trade for White Shark were banned (DOF, 2014b).

In 2007, the NOM-029-PESCA-2006 (NOM-029) was published (DOF, 2007b). This legal instrument provides the specific regulations for shark and ray fisheries in Mexican waters. Some of the conservation measures established in the NOM-059 are the following: (a) a complete ban on shark finning in all fisheries; (b) banning nets from catching sharks by the oceanic industrial fleet since 2009; and (c) prohibition of shark fishing in some critical habitats along the coasts, including nursery areas, and turtle nesting beaches. Shark nursery areas, whether recognized, confirmed (Heupel et al., 2007), or proposed are shown in Fig. 3 in the nursery section below, along with criteria for assigning these designations.

The National Fisheries Chart (CNP) is another of the legally binding instruments used in Mexico to manage shark fisheries (DOF, 2010b). Developed by the National Fisheries and Aquaculture Institute (INAPESCA), the CNP states the status of the stocks, determining the fishing effort allowed to maintain sustainability of associated catches. The CNP also uses other indicators that could be helpful to manage the fishery, including information on climate change factors into some of the fishing stocks (DOF, 2018a).

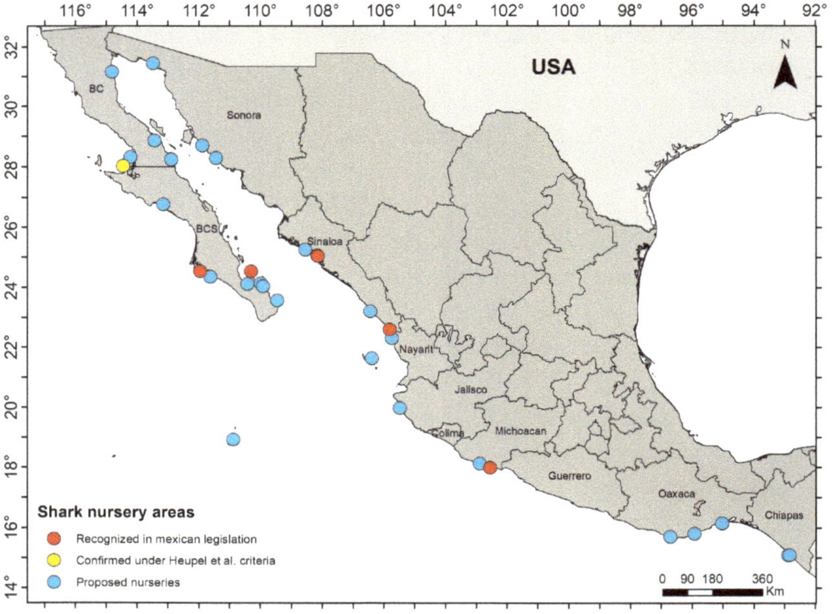


**Fig. 3** Map showing locations of recognized, confirmed, and proposed shark nursery areas in the Mexican Pacific.

Since 2012 the fishing of sharks and rays in the Mexican Pacific has been closed between May 1st and July 31st of every year, with the primary aim to protect the reproductive period of some of the most vulnerable shark and ray species, when they are more susceptible to being caught (DOF, 2012). This prohibition includes the mandatory release of incidental catches of these species. A similar measure has been established in the Mexican waters of the Gulf of Mexico (DOF, 2014c). Furthermore, some Central American countries have established closed fishing seasons similar to that implemented in Mexico, considering that several sharks are transboundary and highly migratory.

### 4.2.3 International and regional management measures

Mexico is a member of two Regional Fisheries Management Organizations (RFMOs) that have adopted important conservation and management measures for sharks: The International Commission for the Conservation of Atlantic Tunas (ICCAT) and the Inter-American Tropical Tuna Commission (IATTC) in the Pacific. Both RFMOs are responsible for the conservation not only of tunas but tuna-like species incidentally caught in tuna fisheries.

In 2005, the IATTC adopted Resolution C-05-03 which placed controls on shark finning by applying a 5% fin-to-carcass weight ratio requirement. The Resolution further resolved that the IATTC would cooperate to provide advice on the status of key shark stocks and propose a research plan for a comprehensive assessment of those stocks (IATTC, 2005). Furthermore, in 2011 the IATTC adopted Resolution C-11-10, which prohibits retaining onboard, trans-shipping, landing, storing, selling, or offering for sale any part or whole carcass of Oceanic Whitetip Sharks, *Carcharhinus longimanus*, in the fisheries covered by the IATTC (IATTC, 2011), and in 2015 Resolution C-15-03 prohibited the intentional setting of purse seine nets for Whale Shark (IATTC, 2015). In 2016, the IATTC adopted Resolution C-16-05, which requires IATTC scientific staff to develop a work plan and timeline for completing stock assessments for hammerhead sharks (Scalloped Hammerhead, Smooth Hammerhead, and Great Hammerhead). Also in 2016 Resolution C-16-04 IATTC called for its Parties, in cooperation with IATTC scientific staff, to research improving the selectivity of fishing gear and improving the knowledge of critical biological and ecological parameters of sharks (including their life-histories, behaviour traits, and migration patterns). This resolution highlighted the need for the Parties to identify key shark mating, pupping, and nursery areas, as well as to improve handling practises for live sharks to maximize their post-release survival (IATTC, 2016a,b,c).

Similarly, the ICCAT has prohibited shark finning and in 2010 adopted some protection measures for Oceanic Whitetip Shark and coastal-oceanic Sphyrna species, prohibiting the retention, trans-shipping, landing, storing, selling, or offering for sale of any part of these species caught in association with fisheries managed by this RFMO. Developing members may still catch these sharks for consumption but must comply with reporting requirements and should not increase their catches (ICCAT, 2010). The ICCAT encourages its Parties to release mako sharks (Shortfin Mako and Longfin Mako) when found alive, to reduce the fishing mortality as the stock status has shown signs of being unhealthy. The ICCAT plans to conduct a future assessment of Shortfin Mako based on Stock Synthesis model that may be more reliable than the previous one based on a production model. To date, both the ICCAT and the IATTC require that catches and discards of sharks are recorded and reported annually at the species level, complemented by observer programmes.

Concern over the status of shark populations around the world, due to the increase of trade and high value of some of their products such as their fins, was stated in the Ninth Conference of the parties of CITES in 1994, of which Mexico is a member. Resolution 9.17 instructed the CITES Committee of Fauna and requested that the FAO and RFMOs gather and review available information about the biology and trade of sharks, to determine those species that might be suitable to be included in the Appendices of the Convention. In 2003, Basking Shark and Whale Shark were included in Appendix II (CITES, 2013), whereas White Shark was added in 2005 (Table 3). In 2014 and 2016 eight more shark species distributed in Pacific Mexico were included: Oceanic Whitetip Shark, Scalloped Hammerhead, Smooth Hammerhead, Great Hammerhead, Silky Shark, and the Thresher Sharks (*Alopias vulpinus, A. pelagicus*, and *A. superciliosus*) (Table 3). All these species have high economic importance for Mexican shark fisheries. As a condition established by CITES, the export of any product of these species requires the issue of an NDF report, which certifies that its trade will not affect its population status.

## 4.3 Challenges for conservation

Due to limited information, few assessments have been undertaken for sharks from the MP, including the full stock assessment of Blue Shark, *P. glauca* (ISC, 2017) and Shortfin Mako, *I. oxyrinchus*, in which information from the Mexican fisheries was used (see below for assessments results). Historical data are largely absent and contemporary data are limited to the period after publication of the NOM-029 in 2007, so rapid evaluation methods like Productivity and Susceptibility Analysis (PSA) (Hobday et al., 2011; Patrick et al., 2010; Walker, 2005) have been used to evaluate the status of some stocks (Benítez et al., 2015; Furlong-Estrada et al., 2014, 2017a; Tovar Ávila et al., 2016).

Historically, shark landings in Mexico were reported into two broad categories: 'tiburón', which included sharks greater than 150 cm in total length (TL), and 'cazón' for sharks less than 150 cm TL (CONAPESCA-INP, 2004). Since the publication of NOM-029 in 2007, the individual catches of some species are reported, although data for some look-alike species, such as hammerhead sharks, are still pooled. This poor taxonomic resolution has limited the available species-specific data need it for the development of legal instruments (e.g. the CNP) and improvement of management measures (e.g. bans, quotes).

Recently, historical catches of Blue Shark in the northern MP were reconstructed using data from the landings of all the fishery units targeting sharks (small-scale, medium, and large vessels or as bycatch in longlines and gillnets) in the states of Baja California, Baja California Sur, Sinaloa, Nayarit, and Colima (Sosa-Nishizaki et al., 2014). Such information, together with the new official information gathered since 2007, has been used in some stock assessments (ISC, 2017).

RFMOs (IATTC and ICATT) also compile fisheries statistics from their members and all entities fishing for tuna in their respective Convention areas, including the MP. They coordinate research (including stock assessments), develop science-based management advice, and provide a mechanism for the Contracting Parties to agree on management measures (IATTC, 2016a,b).

An onboard observers programme (POT) was established in 2006 in the MP industrial fleet as an instrument to collect high-quality fishing and biological data from this fishery (e.g. catch by species, catch per unit effort, size, sex and maturity stage of sharks caught, and survival data). The POT was one of the research priorities mentioned in the NPOA-Mexico and the NOM-029 and it is managed by FIDEMAR (Fideicomiso de Investigación para el desarrollo del Programa Nacional de Aprovechamiento del Atún y Protección de Delfines y Otros en torno a Especies Acuáticas Protegidas), in conjunction with CONAPESCA and INAPESCA, which design the programme, participate in training of the observers, and analyse the information generated. The coverage of the POT has fluctuated over the years, covering an average of 5% of the fishing trips per year, and an expected expansion to at least 20% coverage. The information of the POT is described and has been used in several assessments (Castillo-Géniz et al., 2014; ISC, 2017, 2018; Tovar-Ávila et al., 2011). The development of an onboard observer programme in the semi-artisanal longline shark fishery based in Manzanillo, Colima has also provided data on the species composition of the catches, numerical catch, and CPUE estimates for the 2003–2011 period (Rodríguez-Madrigal et al., 2017b; Santana-Hernández and Flores-Valdez, 2014).

Other efforts have been undertaken to generate high-quality information, including the monitoring of landings of artisanal vessels in the main fishing sites along the whole MP coast by INAPESCA and several universities or research centres like CICESE, CICIMAR, CIBNOR, UAS, UNAM, UMAR, and others. Nevertheless, the large coastal area to cover and limited resources have hampered the continuity of most of the implemented programmes.

### *4.3.1 Populations assessments*

Mexico has participated in two stock assessment undertaken by the International Scientific Committee (ISC) for Tuna and Tuna-like species in the Northern Pacific. First for the Blue Shark, completed in 2017 (ISC, 2017) and second for the Shortfin Mako Shark, completed in 2018 (ISC, 2018). Both stock assessments included standardized CPUE from the MP fishery, based on the information of the POT. Several parameters (sea surface temperature, distance to the coast, year, area fished, and quarter of the year) were analysed with generalized linear models (GLMs) (Fernández-Méndez et al., 2016; González-Ania et al., 2017). The historical trend of for Blue Shark and Shortfin Mako Shark catches from 1976 to 2015 were reconstructed from catch proportions and official statistics for the period 2006–2014 (Sosa-Nishizaki and Castillo-Géniz et al., 2016, Sosa-Nishizaki et al., 2017). Length composition of the catches from both shark species (Castillo-Géniz et al., 2017a,b) and standardized age re-estimations of *I. oxyrinchus* from the MP were also used in the assessment (Rodríguez-Madrigal et al., 2017a).

The results of the stock assessments indicated that the Blue Shark population from the North Pacific is not overfished. Recent decreasing trends in the CPUE of the species in the Eastern Pacific could be explained in terms of recent oceanographic events like the warm Blob of 2013–2015 along the west coast of the northern Mexican Pacific, that could affect the distribution and abundance of Blue Sharks (ISC, 2017).

In the same way, the Shortfin Mako assessment from ISC (2018) indicated that the current spawning abundance relative to unfished levels is 58% (CI = 30–86%; range of 51–68% across alternative scenarios); it is thus unlikely to be overfished. This stock assessment was based in the best available fishing trends for the entire northern hemisphere population (e.g. both the Western and Central as well as the Eastern Pacific), standardized CPUE from 1975 to 2016, size indices from 17 fisheries (including the Mexican Ensenada longline fishery), and biological data integrated in an age-structured population dynamics model constructed in the Stock Synthesis approach. The assessment was undertaken collaboratively by scientists from Canada, Japan, Korea, Mexico, Taiwan, and the United States and was presented to and endorsed by the Western and Central Pacific Fisheries Commission (WCPFC) and the IATTC.

Demographic analysis (life tables, Leslie matrices, and elasticity analysis) have also been used to estimate the resilience of some shark species to fishing mortality rates from the Mexican fisheries, for example, for Silky Shark (Mondragón-Sánchez, 2016) and Shortfin Mako (Mondragón-Sánchez et al., 2018). The preliminary results of both studies have indicated that despite their low productivity, these shark species could attain sustainable catches at current, low fishing mortality levels. This is if fishing focuses only on a fraction of the population, as appears to happen for Shortfin Mako Shark, which is caught by the Mexican fleet only during their early juvenile years. Around 60% of the catches are composed of 2–4-year-old individuals, according to the new growth estimates available (Rodríguez-Madrigal, 2018; Rodríguez-Madrigal et al., 2017a,b).

The PSA evaluations have shown that the risk to shark species of the effects of Mexican shark fisheries ranged from medium to high for the following reasons: (1) their low productivity; and (2) high catch susceptibility, depending on the habits and distribution of the species and the kind of fishery (industrial oceanic or coastal artisanal).

## 4.4 Shark species populations status, possible extirpation of species

### *4.4.1 Population status*

According to the National Fisheries Chart, MP shark fisheries are at their maximum sustainable yield, with 15,000 t of annual catch as a reference point for the period of 1986–2006 (DOF, 2010b). This conclusion is based on the catch trend that includes all shark species because there are no time series of catch and effort by species. The chart also specifies that there has been a change in species composition, where small shark species have replaced large shark species.

Most recent catch records (2000–2017) from the National Commission of Aquaculture and Fisheries (CONAPESCA), show that the average catch is 15,696 t, with the lowest catch in 2001 (11,615 t) and the highest in 2017 (25,385 t). Catches increased gradually from 2000 to 2014, from 11,646 to 14,756, with a significant increase in the last 3 years: 18,006 t in 2015; 21,866 t in 2016; and 25,385 t in 2017. Also, the Fisheries Statistical Yearbook of 2017 indicates that from 2008 to 2017, there was an increase of 7.18% of the global shark catches in Mexico (SAGARPA, 2016). Despite several management measures since 2007, it is difficult to define, with the available information, whether the increase in catches owes to the recovery of some stocks, changes in the dynamics of the fishing fleets, or changes in the composition of species.

The stock assessment for Common Thresher, Blue shark, and Shortfin Mako Shark of the populations shared between the United States and Mexico are robust assessments for shark species in the MP. The evaluations concluded that Common Thresher is neither in an overfished condition nor experiencing overfishing (Teo et al., 2018), the Blue Shark is not overfished (ISC, 2017), and the Shortfin Mako Shark is unlikely to be overfished (ISC, 2018). Other regional assessments, including demographic analysis for Silky Shark and Scalloped Hammerhead in the GOT, concluded that the population of the former was increasing 3–5% per year and the latter decreasing by 6% per year. Note that the entire distribution of the species and multiple fishery indicators are not considered in this evaluation (Soriano-Velásquez et al., 2006). Another study at the entrance of the Gulf of California indicated that Silky Shark and Scalloped Hammerhead appear to be stable in terms of frequency and composition of catches during several decades of fishing exploitation due probably to the restricted distribution of the small-scale fleets and changes in their fishery operations, resulting in low fishing pressure on adults (Furlong-Estrada et al., 2015).

The monitoring of shark catches in the GOC and the GOT carried out by INAPESCA has revealed the decrease in catches of large sharks, such as Bull Shark, Tiger Shark, Lemon Shark, Dusky Shark, and Great Hammerhead (Castillo-Géniz et al., 2008). Many large species that were once common have likely been overfished, prompting a shift in fishing effort towards coastal populations of smaller elasmobranchs (Bizzarro et al., 2009a,b,c). Recently, Saldaña-Ruiz et al. (2017) estimated that in the

GOC, a noticeable decline in landings was observed for Bull Shark, Whitenose Shark, Lemon Shark, *Sphyrna* spp., Bignose Shark, Dusky Shark, Tiger Shark, Pacific Smalltail Shark, Leopard Shark, and Narrowtooth Shark.

Declines in shark catches are also reported in other regions. Interviews with older fishermen in western Baja California suggest that both the abundance and average size of elasmobranchs in catches have declined significantly in recent decades and that, due to the higher ex-vessel prices for most teleost and invertebrate species, a shift in effort to these more sustainable resources is occurring (Cartamil et al., 2011).

### 4.4.2 Possible extirpation of species

The possibility for the local and regional extirpation of some shark species in the MP has been documented. Pérez-Jiménez (2014) found that Scoophead, *Sphyrna media*, Bonnethead, and Great Hammerhead might have been extirpated from the GOC and have become extremely rare in the MTP and GOT. In the last two decades, records of Great Hammerhead (n = 61) in fishery-dependent surveys were restricted to the MTP and GOT, and records of Bonnethead (n = 3) and Scoophead (n = 3) were limited to the Southern region. However, there are 253 catch records of Great Hammerhead, from the period of 2006–2014, in fishery operations of medium-sized vessels in the central-south of the GOC and oceanic waters off north-central Pacific Mexico (Castillo-Géniz et al., 2016); which confirm that this species still occurs in the GOC.

Other species require an analysis of historical records throughout the MP. They are *Ginglymostoma unami*, *C. cerdale*, *C. obscurus*, *N. velox* and *N. brevirostris*. There are very few recent records for these species, in the Gulf of California, only the lemon skark, *N. brevirostris*, has been reproted in the esastern coast of the Baja Peninsula, at Cabo Pulmo (Asúnsolo Rivera, 2016; Bizzarro et al., 2009a,b,c; Pérez-Jiménez et al., 2005). Some of those species have been historically rare in the Gulf of California (Saldaña-Ruiz et al., 2017). Thus, an integral approach that considers an analysis of records in fishery-dependent and independent surveys, ichthyological collections, and the ecological knowledge of fishermen is required to confirm the population decline or regional extirpation of these species.

## 4.5 Transboundary shark species and conservation

### 4.5.1 Transboundary sharks

In the MP there are widely distributed and highly migratory species like Thresher Sharks (*Alopias* spp.), Silky Shark, Blue Shark, and Hammerhead Sharks (*Sphyrna* spp.) (Sosa-Nishizaki et al., 2008). These are among the worldwide recognized commercially important pelagic sharks (Camhi et al., 2009). *Sphyrna lewini* and *C. falciformis* are the species with the highest catches along the Mexican Pacific (Castillo-Géniz et al., 2008; Pérez-Jiménez et al., 2005; Saldaña-Ruiz et al., 2017), except the western coast of Baja California (Cartamil et al., 2011). Both species are highly migratory and transzonal. *Carcharhinus falciformis* has two seasonal movements per year in the Mexican Pacific; the first from north to south (from the Gulf of California to Central America) in the months of March–April, a period in which it is often caught; and the second from south to north in July to September. The migratory movements of *S. lewini* are less well known in Mexican waters (Castillo-Géniz et al., 2008).

The fisheries of highly migratory species are primarily managed by Regional Fisheries Management Organizations (RFMOs). Mexico is part of the regions of the North Pacific Fisheries Commission (NPFC) and the Inter-American Tropical Tuna Commission (IATTC). These organizations have established different recommendations for the management of commercial shark species. These resolutions are related with fishing gears, improvement in the collection of biological-fisheries information, prohibitions, species retention, and commercialization. These recommendations are for the shark species of commercial importance mention in the previous paragraph, and for other highly migratory species like White shark, Basking shark, and Crocodile Shark (Camhi et al., 2009; White et al., 2019).

The international cooperation is crucial for the effective management of high-seas and transboundary shark species. The recommendations issued by the RFMOs are important first steps, but it is necessary more enforceable measures to improve the reports of shark ladings, the bilateral fishery management agreements and the implementation of CITES shark listings based on solid non-detriment findings, when trade in listed species is allowed. The integration of biological and fishery information along the Eastern Pacific is required to conduct comprehensive stock assessments and determine the population status of the shark species.

### 4.5.2 Conservation of sharks

Threatened species are those that could be in danger of disappearing in a period ranging from 5 to 100 years (IUCN, 2017). This is if existing factors that cause the deterioration or modification of their habitat, or directly reduce the size of their populations, do not change substantially in scope and/or scale. Nevertheless, none of the species catalogued as threatened by the NOM-059-SEMARNAT-2010 (Whale Shark, Basking Shark, and White Shark) are considered of commercial importance in Mexican waters (Del Moral-Flores et al., 2016b; DOF, 2010a). Despite this, on the western coast of Baja California, the white shark is incidentally caught in fishery operations of small-scale fleets. For example, Cartamil et al. (2011) recorded from September 2006 to December 2008 the following: (1) the capture of 12 juvenile white sharks in gillnets, representing 0.16% of the elasmobranch catches; and (2) three more juveniles in longline, 0.10% among the catches of elasmobranchs. Additionally, 76 white shark carcasses were found discarded on the beach near the landing site, representing 0.24% of the total recorded elasmobranchs´ carcasses caught in an indeterminate time. The remains of sharks and rays (mainly heads) were discarded next to the fishing camp. In the same area, Castillo-Géniz et al. (2016) recorded in 2015–2015 the catch of 16 juvenile White Sharks in gillnets.

In that same region, specifically in Bahía Sebastián Vizcaino (BSV), Oñate-González et al. (2017) described a breeding area for the White Shark. They state that the zone between BSV to the southern California Bight could be a critical migration corridor for young of the year and juvenile White Sharks. Thus, this region represents an important area for the conservation of this species. Despite not being noted as being of commercial importance, the presence of the Whale Shark is well recognized in the MP. Its movements have been studied in the northeast region (Eckert and Stewart, 2001; Ramírez-Macías et al., 2017) where it is important to ecotourism. By contrast, there are few records of Basking Shark in Mexican waters, with only one confirmed contemporary record in the SCP region in 2003 (Sandoval-Castillo et al., 2008).

## 4.6 Shark nursery areas and their importance for conservation

Nursery areas are essential habitats that protect newborn and juvenile sharks, increasing survival of early life stages and contributing to recruitment to adult populations (Heupel et al., 2007; Kinney and Simpfendorfer, 2009). Successful, persistent nurseries are characterized by high food abundance and favourable environmental conditions, which enables a higher rate of growth and development (Heithaus, 2007). For some species, these areas provide protection from predators, additionally increasing the survival of young sharks. Identifying and protecting nursery areas is essential to improving the conservation of sharks, especially when in conjunction with management actions applied to the older life stages (Heupel et al., 2007; Kinney and Simpfendorfer, 2009).

In the MP, shark catches from artisanal (Bizzarro et al., 2009a,b,c; Cartamil et al., 2011; Pérez-Jiménez et al., 2005; Smith et al., 2009) and industrial fisheries (Castillo-Géniz et al., 2016; Corro Espinosa, 2016; Furlong-Estrada et al., 2017b; García-Rodríguez, 2012; Vögler et al., 2012) are composed mostly of juveniles. Because of that, in Mexican legislation, shark nurseries have been considered as priority areas for shark conservation (DOF, 2007a). The ban on fishing activities inside those areas is a measure to protect these life stages in these habitats (DOF, 2007a).

Mexican resource managers consider the following areas, shown in Fig. 3, as essential for reproduction and parturition (pupping, hatching of sharks in Pacific Mexico, without specifying the species [DOF, 2007a]), and have implemented regulations to protect them:

- Lagoon complex of Bahía Magdalena-Bahia Almejas in Baja California Sur (western BCS),
- Lagoon complex of Bahía Santa María-Bahia Altata in Sinaloa (eastern GOC),
- Coastal areas next to Teacapán, Sinaloa (eastern GOC),
- The coastal line from Río Boca de Campos to Playon de Mexiquillo in Michoacán (Central Mexican Pacific), and
- A radius of 5 km around the seamounts Bajo and Espiritu Santo in Baja California Sur (Southwestern GOC).

In the first four areas the use of gillnets is forbidden in June of every year, while in the last the use of gillnets and longlines is prohibited year-round (DOF, 2007a). Heupel et al. (2007) proposed three elements to identify shark nursery areas: shark density is higher than in other regions; sharks spend an extended period in those habitats; and the same areas are used every year. These criteria have been accepted as the standard to identify nursery areas. Under these criteria, in the MP, only one nursery area has been defined: Bahía Sebastián Vizcaino for the White Shark (Oñate-González et al., 2017). This area has yet to be formally recognized and protected by Mexican legislation. The area around seamounts Bajo and Espiritu Santo in Baja California Sur has been described as a 'hotspot' for subadult females, adult males, and adult females of scalloped hammerhead sharks (Klimley, 1987).

Other areas with high abundance of neonates and juveniles along the coast of the MP have been proposed as nurseries but have not been fully evaluated using the criteria of Heupel et al. (2007). Several authors report that these habitats should be protected for the improvement of shark conservation in the region, complementing those areas recognized in Mexican regulations. Among these areas are:

- The Northern Gulf of California for the Sicklefin Smoothound, *Mustelus lunulatus* and the Grey smoothound, *M. californicus* (Pérez-Jiménez and Sosa-Nishizaki, 2010), as well as the Scalloped Hammerhead, *Sphyrna lewini* and Smooth Hammerhead, *S. zygaena* (Smith et al., 2009).
- Bahia Sebastian Vizcaino for multiple species of sharks (Cartamil et al., 2011), including Blue Shark, *Prionace glauca* and Shortfin Mako, *Isurus oxyrinchus* (Castillo-Géniz et al., 2015).
- Guerrero Negro-Ojo de Liebre, Bahía Magdalena and San Ignacio lagoons for multiple species, including the Swell Shark, *Cephaloscyllium ventriosum* and the Tope Shark, *Galeorhinus galeus* (Ramírez-Amaro et al., 2013).
- Bahia de los Angeles, Bahía Kino and Bahía de la Paz for the Pacific Angel Shark, *Squatina californica* (Villavicencio-Garayzar, 1996).
- Salomón-Aguilar et al. (2009) proposed nursery areas for Pacific Sharpnose Shark, *R. longurio*, Silky Shark, *C. falciformis*, Dusky Shark, *C. obscurus*, Blacktip Shark, *C. limbatus*, and Scalloped Hammerhead, *S. lewini* in San Francisquito and El Barril, Baja California; Bahía Kino and La Manga, Sonora for *C. falciformis*, *C. obscurus*, and *C. limbatus*. Also, describes Bahía Santa María-Bahía Altata and Mazatlán areas in Sinaloa as nurseries for *C. falciformis*, *R. longurio*, *Nasolamia velox*, *S. lewini*, *S. zygaena*, and *M. lunulatus*. These authors further propose the areas of Teacapán Bay, Sinaloa and El Sargento, La Ventana y Punta Arenas in Baja California Sur, as nurseries for many shark species.
- Bahía de la Paz for the Pacific Sharpnose Shark, *Rhizoprionodon longurio* (Trejo-Ramírez, 2017).
- The Revillagigedo and Marias islands for Shortfin Mako, *I. oxyrinchus* (Vélez-Marín and Márquez-Farías, 2009).
- The coast of Sonora and Sinaloa for the Pacific Sharpnose Shark, *R. longurio* (Corro-Espinosa et al., 2011).
- The coast of Sinaloa and Jalisco, Bahía de la Paz and the coast around the mouth of the Nexpa River in Michoacan for Scalloped Hammerhead, *S. lewini* (Anislado-Tolentino, 2000; Rosende-Pereiro and Corgos, 2018; Torres Huerta et al., 2008).
- The coast of Oaxaca (mostly Puerto Escondido and Huatulco Bays) for both Silky Shark, *Carcharhinus falciformis* (Alejo-Plata et al., 2007, 2016; Galván-Tirado et al., 2015; Soriano-Velásquez et al., 2006) and Scalloped Hammerhead, *S. lewini* (Alejo-Plata et al., 2006; Bejarano-Alvarez et al., 2011).

- The Gulf of Tehuantepec for *C. falciformis*, *S. lewini* (Castillo Géniz et al., 2002, 2000) and the Blacktip Shark *C. limbatus* (Soriano-Velásquez et al., 2006).

Future research should focus on confirming if these areas are nurseries and delimiting them into more discrete regions. Also, it is necessary to assess the contribution to recruitment from these areas to adult populations. Data obtained from these studies will help improve understanding of the importance of specific shark nurseries areas and benefit overall shark conservation in the MP.

## 4.7 Protected natural areas

Worldwide, Marine Protected Areas (MPA) are established to increase protection of vulnerable species, conserve biodiversity, and protect critical habitats (IUCN, 2018). Ideally, MPAs are designed to protect all life stages, which is improbable for most shark species due to their high mobility and wide-ranging distribution (Bonfil, 1999; Kinney and Simpfendorfer, 2009). Furthermore, the benefits of MPA protections could be lowered as soon as sharks leave the area and cross into unprotected waters (Bonfil, 1999). In practise, most MPAs are designed to protect sharks in defined life stages, mostly in coastal regions where individuals spend considerable time at vulnerable life stages, like nurseries (Bonfil, 1999; Heupel and Simpfendorfer, 2009; Heupel et al., 2010; Kinney and Simpfendorfer, 2009; Knip et al., 2012). Whether they are small or large, MPAs could improve shark conservation by providing place-based sanctuaries for growth, development, and migration (Heupel and Simpfendorfer, 2009; Knip et al., 2012). In addition to the protection that MPAs provide to sharks, their effectiveness is boosted when coupled with other management strategies in a comprehensive framework (Hillborn, 2017).

In Mexico, Natural Protected Areas (NPA), like MPAs, are established to protect fragile habitats and vulnerable species. These NPAs have two types of areas inside their polygons: core areas and buffer areas. In the first, extractive activities are prohibited but, in the latter, these activities (like fishing) are allowed by local communities (DOF, 1998, 2014d). In the MP there are 21 NPAs decreed that have either an estuarine or a marine portion, from which 10 are Biosphere Reserves, 8 are National Parks (essentially no-take areas), and 3 are Protection Areas of Flora and Fauna (Bonfil, 2014). Most of these areas do not have a management programme that regulates the activities inside the area. From those 21 NPAs, only 7 acknowledge the presence of sharks in their species checklist (Bonfil, 2014), even if their presence has been reported in other documents (Oñate-González et al., 2017). Only three of those areas have regulations that are specific for shark conservation, and these are mostly related to tourism activities (DOF, 2011, 2013, 2018b) (Fig. 4).

Even though some NPAs have been established for more than 20 years, information about the importance of these areas for shark species is lacking. Only an increase of biomass as high as 11 times has been reported for top predators, including Bull Shark, *C. leucas*, Tiger Shark, *Galeocerdo cuvier*, and Whitetip Reef Shark, *Triaenodon obesus*, inside a no-take NPA in the Gulf of California (Aburto-Oropeza et al., 2011). The effectiveness of most of NPAs in the MP has not been evaluated, so it is not clear if their establishment is improving shark conservation in any meaningful way.

Mexican regulations for sharks establish that essential habitats for this group should be protected, so nursery areas have been proposed as potential NPAs in the MP (Salomón-Aguilar et al., 2009). Besides the creation of new NPAs that contribute to shark conservation (based on robust information of their importance), future research should focus on verifying if existing NPAs are improving conservation benefits to this group and put this information into specific regulations inside management programmes for each NPA.

A reserve has been created consisting of 57,000 mile$^2$ around the Revillagigedo Archipelago, a UNESCO World Heritage site composed of a string of four volcanic islands about 300 miles southwest of the Baja Peninsula (DOF, 2018b). The designation controls mining, fishing and tourism development on or near the islands. This protects key habitat for six species of sharks (Silvertip Shark, Whitetip reef shark, Tiger shark, Silky shark, Galapagos Shark, and Scalloped Hammerhead) (DOF, 2018b). It used to be protected 6 miles around every island but via acoustic tracking sharks connectivity was established between all these islands, resulting in protection 40 mile$^2$ around the islands. The first design was proposed to the Mexican Commission of NPAs in 2014 and included in the UNESCO Heritage site documentation. Another, more expanded, design was used by a group of lawyers in Mexico City and Pew Charitable Trusts to push for the creation of the Revillagigedo National Park.

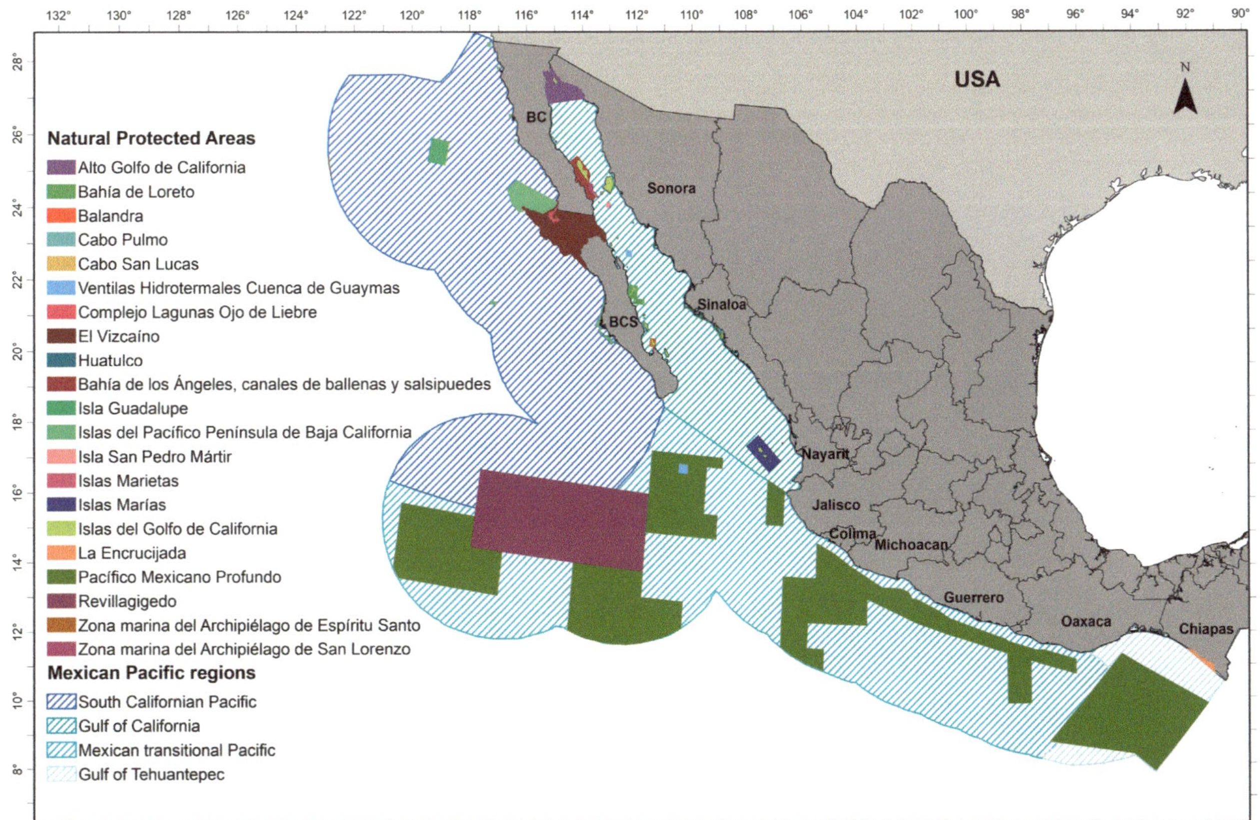


**Fig. 4** Delimitation of Natural Protected Areas in the Mexican Pacific. *Data from CONANP, 2019 Comisión Nacional de Áreas Naturales Protegidas. (Julio 2019). Mapas de regiones CONANP. Retrieved from http://sig.conanp.gob.mx/website/pagsig/info_shape.htm.*

## 5. Conclusions

Despite the ecological and economic importance of sharks, relatively little is known about their biodiversity and distribution, and there are still substantial gaps in the biological information for many species, in the MP. The systematic listings of species and the ichthyological collections generated over the years are vital for understanding biodiversity and it is through these data that we can begin to elucidate the complete panorama of the distribution of these species in the MP, and its relation to physical and chemical parameters. Much work, however, remains to be done before the full ecological role of this group can be appreciated and protected.

The physical environment modulates the distribution of shark species in the MP, and the specific conditions of each region of the MP carry with it a distinct catalogue of species. The influence south of the California Current and North Equatorial Counter Current is observed in differences in the prevalence and predominance of species between northern (subtropical species) and southern (tropical species) regions. The knowledge of distribution patterns is essential to understanding the role of sharks in the MP ecosystems, to elucidate the shifts in distribution and changes in the timing of migrations of the species in response to climate change, for a better understanding of the overlap of species with fishing gears and fleet, and as a reference for the effective conservation of their populations.

Much of the information about the presence and description of species in different areas of the MP comes from fishery-dependent sampling, which creates a bias since only target species of commercial importance are adequately considered. In the MP shark fishing is very important, with a long history and tradition in Mexico. There is a keen demand for shark meat in Mexico, and for that reason it is very important to modulate fishing effort based on knowledge of the status of the population to develop better management and conservation plans.

Most of the MP exclusive economic zone consists of depths greater than 2000 m. Despite this, the research of deep waters within the MP is also limited—very few studies have focused on this area and its unique fauna. For all the above reasons, research efforts should focus on fully documenting the biodiversity of shark species throughout the exclusive economic zone regardless of their commercial value. Another fundamental need for understanding the biodiversity of sharks in the MP is accurate cataloguing and description of the species present. A taxonomic and systematic revision of the species described should be a priority to resolve discrepancies between lists of species due to current taxonomic adjustments.

The scarcity of unbiased information on shark species in the MP is a serious problem that makes it impossible to carry out data-rich species assessments. The lack of data also interferes with the correct application of CITES measures. The issuance of the NDF is one of the main problems of the CITES authority in Mexico. Since there are still information gaps to be filled, there is no adequate monitoring of the species. The communications channels between government agencies have to be improved, and the trade records of shark species listed in CITES remain mostly unknown. Finally, there is an increasing concern for the Threatened status of sharks in the MP. Nearly 30% of the MP shark species are Threatened, and another 30% have the designation of Data Deficient. Given the possibility of extirpation of several species of the Sphyrna genus, and the reduction of records for many species that were formerly more common, the monitoring of shark biodiversity in the MP should be a conservation priority. Most marine conservation has focused on protecting some charismatic species or on managing species of fishing and economic importance. However, the enormous biodiversity of sharks in the MP described in this chapter indicates the need for studies directed towards other species and their relationships with the ecosystem.

## Acknowledgements

We want to thank our colleagues Peter Klimley, Dayv Lowry, and one anonymous reviewer for their editorial efforts and for taking the time and effort to review this chapter carefully. We sincerely appreciate all the valuable comments and suggestions, which helped us to improve the quality of the manuscript. We also thank Zurisaday Ramírez Mendoza for their help with de data for the maps. We also thank Oscar Sosa-Nishizaki, who supported the participation of Emiliano García Rodríguez in this chapter, and for their support during the creation of this chapter.

## Appendix. Checklist of Pacific Mexico sharks

Checklist of Pacific Mexico sharks. Scientific name, common name in English and Spanish, distribution of each species by region of the Mexican Pacific, habitat, and IUCN status are shown.

| Order | Family | Scientific name | Common name | Pacific Mexico region | Habitat | IUCN status |
|---|---|---|---|---|---|---|
| Heterodontiformes | Heterodontidae | *Heterodontus francisci* | Horn shark/Tiburón puerco | SCP, GOC | Coastal | Data Deficient |
| | | *Heterodontus mexicanus* | Mexican Horn Shark/ Tiburón perro | GOC, MTP, GOT | Coastal | Data Deficient |
| Orectolobiformes | Ginglymostomatidae | *Ginglymostoma unami* | Pacific Nurse Shark/ Tiburón gata | SCP, GOC, MTP, GOT | Coastal | Not Evaluated |
| | Rhincodontidae | *Rhincodon typus* | Whale Shark/Tiburón ballena | SCP, GOC, MTP, GOT | Coastal/ Oceanic | Endangered* |
| Lamniformes | Odontaspididae | *Odontaspis ferox* | Smalltooth Sand Tiger/ Tiburón dientes de perro | SCP, GOC | Deepwater | Vulnerable |
| | | *Odontaspis noronhai* | Bigeye Sand Tiger Shark/ Solrayo ojigrande | SCP, GOC, MTP, GOT | Deepwater | Data Deficient |
| | Pseudocarchariidae | *Pseudocarcharias kamoharai* | Crocodile Shark/Tiburón cocodrilo | SCP | Oceanic | Near Threatened |
| | Alopiidae | *Alopias pelagicus* | Pelagic Thresher/Zorro pelágico | SCP, GOC, MTP, GOT | Oceanic | Vulnerable |
| | | *Alopias superciliosus* | Bigeye Thresher/Tiburón zorro ojón | SCP, GOC, MTP, GOT | Oceanic | Vulnerable |
| | | *Alopias vulpinus* | Common Thresher Shark/ Tiburón zorro común | SCP, GOC, MTP, GOT | Coastal/ Oceanic | Vulnerable |
| | Megachasmidae | *Megachasma pelagios* | Megamouth Shark/Tiburón bocon | SCP, GOC | Coastal/ Oceanic | Least Concern |
| | Cetorhinidae | *Cetorhinus maximus* | Basking shark/Tiburón peregrino | SCP, GOC, MTP, GOT | Oceanic | Vulnerable* |

*Continued*

—cont'd

| Order | Family | Scientific name | Common name | Pacific Mexico region | Habitat | IUCN status |
|---|---|---|---|---|---|---|
| | Lamnidae | *Carcharodon carcharias* | White shark/Tiburón blanco | SCP, GOC | Semi-oceanic | Vulnerable* |
| | | *Isurus oxyrinchus* | Shortfin mako/Tiburón mako | SCP, GOC, MTP, GOT | Coastal/Oceanic | Vulnerable |
| | | *Isurus paucus* | Longfin Mako/Tiburón aleton | SCP | Deepwater | Vulnerable |
| | | *Lamna ditropis* | Salmon shark/Tiburón salmon | SCP | Coastal/Oceanic | Least Concern |
| Carcharhiniformes | Pentanchidae | *Apristurus brunneus* | Brown Cat Shark/Pejegato marrón | SCP | Deepwater | Data Deficient |
| | | *Apristurus kampae* | Longnose Catshark/Pintarroja | SCP, GOC | Deepwater | Data Deficient |
| | | *Apristurus nasutus* | Largenose Catshark/Tiburón narizón | GOC | Deepwater | Data Deficient |
| | | *Cephalurus cephalus* | Lollipop Catshark/Tiburón renacuajo | GOC | Deepwater | Data Deficient |
| | | *Galeus piperatus* | Peppered Catshark/Colayo punteado | GOC | Deepwater | Least Concern |
| | | *Parmaturus xaniurus* | Filetail Cat Shark/Pejegato lima | SPC,GOC | Deepwater | Least Concern |
| | Scyliorhinidae | *Cephaloscyllium ventriosum* | Swell Shark/Pejegato globo | SCP, GOC, MTP, GOT | Coastal | Least Concern |
| | Triakidae | *Galeorhinus galeus* | Tope Shark/Tiburón aceitoso | SCP, GOC | Coastal | Vulnerable |
| | | *Mustelus albipinnis* | Whitemargin Smoothhound/Cazón de hacat | GOC | Deepwater | Data Deficient |

| | | | | | | |
|---|---|---|---|---|---|---|
| | | *Mustelus californicus* | Grey smoothhound/Cazon mamón | SPC,GOC | Coastal | Least Concern |
| | | *Mustelus dorsalis* | Sharptooth smoothhound/Tiburón mamón enano | GOC, MTP, GOT | Coastal | Data Deficient |
| | | *Mustelus henlei* | Brown smoothhound/Cazón hilacho | SCP, GOC | Coastal | Least Concern |
| | | *Mustelus lunulatus* | Sicklefin Smoothhound/Cazon segador | SCP, GOC, MTP, GOT | Coastal | Least Concern |
| | | *Triakis semifasciata* | Leopard shark/Tiburón leopardo | SPC,GOC | Coastal | Least Concern |
| | Carcharhinidae | *Carcharhinus albimarginatus* | Silvertip Shark/Tiburón puntas blancas | SCP | Coastal | Vulnerable |
| | | *Carcharhinus altimus* | Bignose Shark/Tiburón narizón | SCP, GOC, MTP | Coastal | Data Deficient |
| | | *Carcharhinus brachyurus* | Narrowtooth shark/Tiburón cobrizo | SCP, GOC | Coastal | Near Threatened |
| | | *Carcharhinus cerdale* | Pacific Smalltail Shark/Tiburón poroso del pacífico | SCP, GOC, MTP, GOT | Coastal | Not Evaluated |
| | | *Carcharhinus falciformis* | Silky Shark/Tiburón piloto | SCP, GOC, MTP, GOT | Oceanic | Vulnerable |
| | | *Carcharhinus galapagensis* | Galapagos Shark/Tiburón de Galápagos | SCP, GOC | Coastal | Near Threatened |
| | | *Carcharhinus leucas* | Bull Shark/Tiburón toro | SCP, GOC, MTP, GOT | Coastal | Near Threatened |
| | | *Carcharhinus limbatus* | Blacktip Shark/Tiburón volador | SCP, GOC, MTP, GOT | Coastal | Least Concern |

*Continued*

—cont'd

| Order | Family | Scientific name | Common name | Pacific Mexico region | Habitat | IUCN status |
|---|---|---|---|---|---|---|
| | | *Carcharhinus longimanus* | Oceanic Whitetip Shark/ Tiburón oceánico | SCP, GOC, MTP, GOT | Oceanic | Vulnerable |
| | | *Carcharhinus obscurus* | Dusky Shark/Tiburón gambuso | SCP, GOC | Coastal | Vulnerable |
| | | *Carcharhinus porosus* | Smalltail Shark/Tiburón poroso | GOC, MTP, GOT | Coastal | Data Deficient |
| | | *Galeocerdo cuvier* | Tiger Shark/Tintorera | SCP, GOC, MTP, GOT | Coastal | Near Threatened |
| | | *Nasolamia velox* | Whitenose Shark/Tiburón coyotito | SCP, GOC, MTP, GOT | Coastal | Data Deficient |
| | | *Negaprion brevirostris* | Lemon Shark/Tiburón limón | SCP, GOC, MTP, GOT | Coastal | Near Threatened |
| | | *Prionace glauca* | Blue shark/Tiburón azul | SCP, GOC, MTP, GOT | Oceanic | Near Threatened |
| | | *Rhizoprionodon longurio* | Pacific Sharpnose Shark/ Cazón bironche | SCP, GOC, MTP, GOT | Coastal | Data Deficient |
| | | *Triaenodon obesus* | Whitetip reef shark/Cazón coralero trompacorta | MTP | Coastal | Near Threatened |
| | Sphyrnidae | *Sphyrna corona* | Scalloped Bonnethead/ Cornuda coronada | GOC, MTP, GOT | Coastal | Near Threatened |
| | | *Sphyrna lewini* | Scalloped Hammerhead/ Cornuda común | SCP, GOC, MTP, GOT | Semi-oceanic | Endangered |
| | | *Sphyrna media* | Scoophead/Cornuda cuchara | GOC, MTP, GOT | Coastal | Data Deficient |
| | | *Sphyrna mokarran* | Great Hammerhead/ Cornuda gigante | SCP, GOC, MTP, GOT | Semi-oceanic | Endangered |

| | | | | | | |
|---|---|---|---|---|---|---|
| | | *Sphyrna tiburo* | Bonnethead/Cornuda cabeza de pala | SCP, GOC, MTP, GOT | Coastal | Least Concern |
| | | *Sphyrna zygaena* | Smooth Hammerhead/ Cornuda prieta | SCP, GOC | Semi-oceanic | Vulnerable |
| Hexanchiformes | Chlamydoselachidae | *Chlamydoselachus anguineus* | Frilled shark/Tiburón anguila | SCP | Deepwater | Least Concern |
| | Hexanchidae | *Hexanchus griseus* | Bluntnose Sixgill Shark/ Tiburón de seis branquias | SPC,GOC | Deepwater | Near Threatened |
| | | *Notorynchus cepedianus* | Broadnose Sevengill Shark/ Tiburón pinto | SPC,GOC | Coastal | Data Deficient |
| Squaliformes | Dalatiidae | *Euprotomicrus bispinatus* | Pygmy Shark/Tiburón pigmeo | SCP | Deepwater | Least Concern |
| | | *Isistius brasiliensis* | Cookiecutter Shark/ Tiburón cigarro | SCP | Deepwater | Least Concern |
| | Etmopteridae | *Centroscyllium nigrum* | Combtooth Dogfish/Tollo negro | SCP | Deepwater | Data Deficient |
| | Somniosidae | *Somniosus pacificus* | Pacific Sleeper Shark/ Tiburón dormilón del Pacífco | SCP | Deepwater | Data Deficient |
| | Squalidae | *Squalus suckleyi* | North Pacific Spiny Dogfish/Cazón espinoso común | SPC,GOC | Deepwater | Least Concern |
| Echinorhiniformes | Echinorhinidae | *Echinorhinus cookei* | Prickly Shark/Tiburon espinozo negro | SCP, GOC, MTP, GOT | Deepwater | Data Deficient |
| Squatiniformes | Squatinidae | *Squatina californica* | Pacific Angel Shark/ Angelote del pacifico | SCP, GOC | Coastal | Near Threatened |

The systematic arrangement is based on Nelson et al. (2016).
*These species are also listed as Threatened by Mexican regulations and its fishing is prohibited in Mexican waters (DOF, 2014b).

## References


Abitia-Cárdenas, L.A., Rodríguez–Romero, J., Galván–Magaña, F., De La Cruz–Agüero, J., Chávez–Ramos, H., 1994. Lista sistemática de la ictiofauna de Bahía de La Paz, Baja California Sur, Mexico. Cienc. Mar. 20 (2), 159–181.

Aburto-Oropeza, O., Erisman, B., Galland, G.R., Mascareñas-Osorio, I., Sala, E., Ezcurra, E., 2011. Large recovery of fish biomass in a no-take marine reserve. PLoS One 6 e23601.

Acevedo-Cervantes, A., López–Martínez, J., Herrera–Valdivia, E., Rodríguez–Romero, J., 2009. Nuevo registro de tiburón narizón, *Apristurus nasutus* de Buen (Carcharhiniformes: Scyliorhinidae), en el Golfo de California, Mexico. Hidrobiológica 19 (3), 269–272.

Aguirre-Villaseñor, H., Salas-Singh, C., 2012. New records of the lollipop catshark *Cephalurus cephalus* (Scyliorhinidae) from the Gulf of California, Mexico. Rev. Mex. Biodiv. 83 (1), 298–300.

Alejo-Plata, M.D.C., Ramos-Carrillo, S., Cruz-Ruiz, J.L., 2006. La pesquería artesanal del tiburón en Salina Cruz, Oaxaca, México. Cienc. Mar. X, 37–51.

Alejo-Plata, M.D.C., Gómez-Márquez, J.L., Ramos, S., Herrera, E., 2007. Presencia de neonatos y juveniles del tiburón martillo *Sphyrna lewini* (Griffith & Smith, 1834) y del tiburón sedoso *Carcharhinus falciformis* (Müller & Henle, 1839) en la costa de Oaxaca, Mexico. Rev. Biol. Mar. Ocean. 42, 403–413.

Alejo-Plata, M.D.C., Ahumada-Sempoal, M.A., Gómez-Márquez, J.L., González-Acosta, A., 2016. Population structure and reproductive characteristics of the silky shark *Carcharhinus falciformis* (Müller & Henle, 1839) (Carcharhiniformes: Carcharhinidae) off the coast of Oaxaca, Mexico. Lat. Am. J. Aquat. Res. 44, 513–524.

Álvarez-Borrego, S., 2010. Physical, chemical, and biological oceanography of the Gulf of California. In: Brusca, R.C. (Ed.), The Gulf of California: Biodiversity and Conservation. In: vol. 1. University of Arizona Press, Arizona, pp. 24–48.

Álvarez-León, R., Castro-Aguirre, J.L., 1983. Notas sobre la captura incidental de dos especies de tiburón en las costas de Mazatlán (Sinaloa) Mexico. Stud. Neotropical Fauna Environ. 18 (4), 201–207.

Álvarez-Romero, J.G., Presseya, R.L., Bana, N.C., Torre-Cosío, J., Aburto-Oropeza, O., 2013. Marine conservation planning in practice: lessons learned from the Gulf of California. Aquat. Conserv. Mar. Freshwat. Ecosyst. 23, 483–505.

Anislado-Tolentino, V., 2000. Ecología pesquera del tiburón martillo *Sphyrna lewini* (Griffith y Smith, 1834) en el litoral del estado de Michoacán, Mexico. Unpublished Master's thesis, UNAM, México.

Applegate, S.P., Espinosa, L., Menchaca, L.B., Sotelo, F., 1979. Tiburones Mexicanos. Subsecretaría de Educación e Investigación Tecnológica. Dirección General de Ciencia y Tecnología del Mar, México.

Asúnsolo Rivera, A., 2016. Distribución, Abundancia Y Riqueza de Especies de Tiburones del Arrecife de Cabo Pulmo, B.C.S. México. Unpublished Undergraduate Thesis, Universidad Autónoma de Baja California Sur, Mexico.

Badan, A., 1997. La Corriente Costera de Costa Rica en el Pacífico mexicano. Monogr. No. 3. Unión Geofísica Mexicana, pp. 99–112.

Balart, E.F., González-García, J., Villavicencio-Garayzar, C., 2000. Notes on the biology of *Cephalurus cephalus* and *Parmaturus xaniurus* (Chondrichthyes: Scyliorhinidae) from the west coast of Baja California Sur, Mexico. Fish. Bull. 98, 219–221.

Bejarano-Alvarez, M., Galván-Magaña, F., Ochoa-Baez, R.I., 2011. Reproductive biology of the scalloped hammerhead shark *Sphyrna lewini* (Chondrichthyes: Sphyrnidae) off South-West Mexico. Aqua Int. J. Ichthyol. 17, 1–15.

Benítez, H., López, G., Rivera-Téllez, E., 2015. Taller de Evaluación de Productividad, Susceptibilidad y Manejo de tiburones mexicanos listados en el Apéndice II de la CITES. Informe de Resultados. Comisión Nacional Para el Conocimiento y Uso de la Biodiversida, Mexico, México, DF.

Bizzarro, J.J., Smith, W.D., Hueter, R.E., Villavicencio-Garayzar, C.J., 2009a. Activities and catch composition of artisanal elasmobranch fishing sites on the eastern coast of Baja California Sur, Mexico. Bull. South Calif. Acad. Sci. 108 (3), 137–151.
Bizzarro, J.J., Smith, W.D., Márquez–Farías, J.F., Tyminski, J., Hueter, R.E., 2009b. Temporal variation in the artisanal elasmobranch fishery of Sonora, Mexico. Fish. Res. 97 (1–2), 103–117.
Bizzarro, J.J., Smith, W.D., Castillo–Géniz, J.L., Ocampo–Torres, A., Márquez–Farías, J.F., Hueter, R.E., 2009c. The seasonal importance of small coastal sharks and rays in the artisanal elasmobranch fishery of Sinaloa, México. Pan Am. J. Aquat. Sci. 4 (4), 513–531.
Bonfil, R., 1999. Marine protected areas as a shark fisheries management tool. In: Seret, B., Sire, J.Y. (Eds.), Proceedings of the 5th Indo-Pacific Fish Conference Noumea, 1997. Societe Francaise d'Ichtyologie. Paris and Institut de Recherche pour le Development, Paris, pp. 217–230.
Bonfil, R., 2014. Diagnóstico del Estado de Conservación de los Elasmobranquios en México. Informe final de la consultoría, p. 109.
Brusca, R.C., Findley, L.T., Hastings, P.A., Hendrickx, M.E., Cosio, J.T., van der Heiden, A.M., 2005. Macrofaunal diversity in the Gulf of California. In: Cartron, J.L.E., Ceballos, G., Felger, R.S. (Eds.), Biodiversity, Ecosystems, and Conservation in Northern Mexico. Oxford University Press, New York, pp. 179–203.
Buckhorn, M.L., 2012. Guide to the Marine Fishes of the Gulf of California. American Fisheries Society, Bethesda, Maryland.
Cailliet, G.M., Musick, J.A., Simpfendorfer, G.A., Stevens, J.D., 2005. Ecology and life history characteristics of chondrichthyan fish. In: Fowler, S.L., Cavanagh, R.D., Camhi, M., Burgess, G.H., Cailliet, G.M., Fordham, S.V., Simpfendorfer, C.A., Musick, J.A. (Eds.), Sharks, Rays and Chimaeras: the Status of the Chondrichthyan Fishes: Status Survey and Conservation action Plan. IUCN SSC Shark Specialist Group, Gland, Switzerland, pp. 12–16.
Camhi, M.D., Valenti, S.V., Fordham, S.V., Fowler, S.L., Gibson, C., 2009. The Conservation Status of Pelagic Sharks and Rays: Report of the IUCN Shark Specialist Group Pelagic Shark Red List Workshop. IUCN Species Survival Commission Shark Specialist Group, Newbury, UK.
Carrier, J.C., Musick, J.A., Heithaus, M.R., 2010. Sharks and Their Relatives II: Biodiversity, Adaptive Physiology, and Conservation. CRC Press, New York.
Cartamil, D., Santana-Morales, O., Escobedo-Olvera, M., Kacev, D., Castillo-Géniz, L., Graham, J.B., Rubin, R.D., Sosa-Nishizaki, O., 2011. The artisanal elasmobranch fishery of the Pacific coast of Baja California, Mexico. Fish. Res. 108 (2), 393–403.
Castillo Géniz, J.L., Soriano Velásquez, S.R., Sancho Vásquez, F., Ramírez Santiago, C.E., Cid del Prado Vera, A., 2002. Elasmobranquios del Golfo de Tehuantepec, litoral chiapaneco. Universidad Autónoma de Baja California Sur. Informe final SNIB-CONABIO proyecto No. S123, México.
Castillo-Géniz, J.L., Márquez-Farías, J.F., Cid del Prado-Vera, A., Soriano Velásquez, S.R., Corro-Espinosa, D., Ramírez-Santiago, C., 2000. Tiburones del Pacífico mexicano. Pesquería artisanal. In: INP, (Ed.), Sustentabilidad y Pesca Responsable en México: Evaluación y Manejo 1997–1998. Instituto Nacional de la Pesca, Secretaría de Medio Ambiente, Recursos Naturales y Pesca, pp. 127–151.
Castillo-Géniz, J.L., Soriano-Velázquez, S.R., Villaseñor-Talavera, R., 2008. Pesquerías Mexicanas de tiburón en el Oceáno Pacífico. In: Machii, T., Flores, O.J. (Eds.), Pesquerías Latinoamericanas. Comisión Nacional de Acuacultura y Pesca. Secretaría de Agricultura, Ganadería, Desarrollo Rural, Pesca y Alimentación, México, pp. 211–241.
Castillo-Géniz, J.L., Ocampo–Torres, A.I., Shimada, K., Rigsby, C.K., Nicholas, A.C., 2012. Tiburón bocudo juvenil, *Megachasma pelagios*, capturado en la costa del Pacífco de México, y su relevancia para la diversidad de los peces condrictios en México. Cienc. Mar. 38 (2), 467–474.

Castillo-Géniz, J.L., Gódinez-Padilla, C.J., Ajás-Terriquez, H.A., González-Ania, L.V., 2014. Catch data for shorfin mako Shark reported by fishery observers from Mexican Shark longline and driftnet fisheries in the North Pacific in 2006–2014. ISC Shark Working Group Workshop. ISC/14/SHARKWG-3/02.
Castillo-Géniz, J.L., Godínez Padilla, C.J., Ortega-Salgado, I., Ajás-Terriquez, H.A., González-Ania, L.V., Fernández-Méndez, J.I., Herrera-Gómez, M.E., Pérez-Weil, J.R., 2015. Estudios sobre pesquerías de tiburones y rayas del noroeste del Pacífico mexicano: investigaciones aplicadas para el mejoramiento de su manejo pesquero y conservación. Technical report INAPESCA- CRIP.
Castillo-Géniz, J.L., Godinez Padilla, C.J., Ortega Salgado, I., Ajas Terriquez, H.A., 2016. La importancia Pesquera de los tiburones incluidos en el Apendice II de la CITES en aguas de Mexico: Programa de observadores de tiburón. In: Castillo-Géniz, J.L., Tovar Avila, J. (Eds.), Tiburones mexicanos de importancia Pesquera en la CITES. Instituto Nacional de Pesca, Mexico, pp. 29–36.
Castillo-Géniz, J.L., Godínez-Padilla, C.J., González-Ania, L.V., Haro-Avalos, H., Mondragón-Sánchez, L.F., Tovar-Avila, J., 2017a. Size and sex of the Blue Sharks caught by the Mexican longline industrial fleets recorded by on board Observers in the Pacific 2006–2015. ISC Shark Working Group Workshop. ISC/17/SHARKWG-1/01.
Castillo-Géniz, J.L., Godínez-Padilla, C.J., González-Ania, L.V., Haro-Avalos, H., Fernández-Méndez, J.I., 2017b. Size and sex of the shortfin mako caught by the Mexican longline industrial fleets recorded by on board observers in the Pacific 2006–2016. ISC Shark Working Group Workshop. ISC/17/SHARKWG-3/11.
Castro, J.I., 2011a. The Sharks of North America. Oxford University Press, New York.
Castro, J.I., 2011b. Resurrection of the name *Carcharhinus cerdale*, a species different from *Carcharhinus porosus*. Aqua Int. J. Ichthyol. 17, 1), 1–11.
Castro, R., Collins, C.A., Rago, T.A., Margolina, T., Navarro-Olache, L.F., 2017. Currents, transport, and thermohaline variability at the entrance to the Gulf of California (19–21 April 2013). Cienc. Mar. 43 (3), 173–190.
Castro-Aguirre, J.L., 1981. Especies de la familia Scyliorhinidae (Elasmobranchii, Galeoidea), de la costa occidental de México, con especial referencia a *Cephalurus cephalus* (Gilbert). An. Esc. Nac. Cienc. Biol. 24 (1–4), 71–93. México.
Castro-Aguirre, J.L.C., Balart, E.F., 1996. Contribución al conocimiento del origen y relaciones de la ictiofauna de aguas profundas del Golfo de California, México. Hidrobiológica 6 (1–2), 67–76.
Castro-Aguirre, J.L., García-Domínguez, F., 1988. Sobre la presencia de *Isistius brasiliensis* (Quoy y Gaimard) (Squaliformes: Squalidae: Dalatiinae) en el Golfo de México. Con un elenco sistemático de las especies mexicanas pertenecientes al superorden Squalomorphii. An. Esc. Nac. Cienc. Biol. 32, 91–108. México City.
Castro-Aguirre, J.L., Antuna–Mendiola, A., González–Acosta, A.F., De La Cruz–Agüero, J., 2003. Primer registro del tiburón cañabota *Hexanchus griseus* (Bonnaterre, 1788) (Chondrichthyes: Selachiimorpha) en la costa occidentalde Baja California Sur, México. Oceánides 18 (1), 39–41.
CEC (Commission for Environmental Cooperation), 2017. Sustainable Trade in Sharks: Action Plan for North America. Commission for Environmental Cooperation, Montreal, Canada, p. 56.
CITES, 2013. Notification to the Parties No. 2013/012: Amendments to Appendices I and II of the Convention. CITES Secretariat, Geneva.
Compagno, L.J., 2008. Pelagic elasmobranch diversity. In: Camhi, M.D., Pikitch, E.K., Babcock, E.A. (Eds.), Sharks of the Open Ocean: Biology, Fisheries and Conservation. In: 15John Wiley & Sons.
CONAPESCA-INP, 2004. Plan de Acción Nacional para el Manejo y Conservación de Tiburones, Rayas y Especies Afines en México. SAGARPA, México. 85 p.

Corro Espinosa, D., 2016. Sinaloa. In: Castillo Géniz, J.L., Castillo-Géniz, J.L., Tovar Ávila, J. (Eds.), Tiburones mexicanos de importancia Pesquera en la CITES. Instituto Nacional de Pesca, Mexico, pp. 39–46.
Corro-Espinosa, D., Márquez-Farías, J.F., Muhlia-Melo, A., 2011. Size at maturity of the Pacific Sharpnose shark *Rhizoprionodon longurio* in the Gulf of California, Mexico. Cienc. Mar. 37 (2), 201–214.
Cortés-Fuentes, C., 2018. Distribución espacio-temporal y preferencias ambientales de tiburones en el área natural protegida Archipiélago de Revillagigedo. Unpublished Master's thesis, CIBNOR, México.
Cruz, A., Soriano, S.R., Santana, H., Ramírez, C.E., Valdez, J.J., 2011. La pesquería de tiburones oceánicos-costeros en los litorales de Colima, Jalisco y Michoacán. Rev. Biol. Trop. 59 (2), 655–667.
De La Cruz-Agüero, J., Arellano–Martínez, M., Cota–Gómez, V.M., 1996. Lista sistematica de los peces marinos de las Lagunas Ojo de Liebre y Guerrero Negro, BCS y BC, México. Cienc. Mar. 22 (1), 111–128.
De La Cruz-Agüero, J., Arellano-Martínez, M., Cota-Gómez, V.M., De La Cruz-Agüero, G., 1997. Catálogo de los peces marinos de Baja California Sur. Centro Interdisciplinario de Ciencias Marinas, IPN and Comisión Nacional para el Conocimiento y Uso de la Biodiversidad. Ciudad de México.
De la Lanza, G.E., 1991. Oceanografía de Mares Mexicanos. AGT Editor, México.
Del Moral-Flores, L.F., Morrone, J.J., Alcocer Durand, J., Espinosa-Pérez, H., Pérez-Ponce de León, G., 2015a. Lista patrón de los tiburones, rayas y quimeras (Chondrichthyes, Elasmobranchii, Holocephali) de México. Arx. Misc. Zool. 13, 47–163.
Del Moral-Flores, L.F., Ramírez-Antonio, E., Angulo, A., Pérez-Ponce de León, G., 2015b. Ginglymostoma unami sp. nov. (Chondrichthyes: Orectolobiformes: Ginglymostomatidae): una especie nueva de tiburón gata del Pacífco oriental tropical. Rev. Mex. Biodiv. 86, 48–58.
Del Moral-Flores, L.F., Morrone, J.J., Alcocer Durand, J., Ponce de León, G., 2016a. Diversidad y afnidades biogeográfcas de los tiburones, rayas y quimeras (Chondrichthyes: Elasmobranchii, Holocephali) de México. Rev. Biol. Trop. 64, 1469–1486.
Del Moral-Flores, L.F., Anislado-Tolentino, V., Martínez-Ramírez, E., Pérez-Ponce de León, G., Ramírez-Antonio, E., González-Medina, G., 2016b. Ictiofauna marina de Oaxaca, México: listado sistemático y afnidades zoogeográfcas. Acta Univ. 27 (1), 3–25.
DOF, Diario Oficial de los Estados Unidos Mexicanos, 1998. Ley General del Equilibrio Ecológico y la Protección al Ambiente. Secretaría de Medio Ambiente y Recursos Naturales, México D.F.
DOF, Diario Oficial de los Estados Unidos Mexicanos, 2007a. Ley General de Pesca y Acuicultura. Secretaría de Medio Ambiente y Recursos Naturales, México D.F.
DOF, Diario Oficial de los Estados Unidos Mexicanos, 2007b. Norma Oficial Mexicana NOM-029-PESC-2006, Pesca responsable de tiburones y rayas. Especificaciones para su aprovechamiento. Secretaría de Medio Ambiente y Recursos Naturales, México D.F.
DOF, Diario Oficial de los Estados Unidos Mexicanos, 2010a. Norma Oficial Mexicana NOM-059-SEMARNAT-2010. Protección ambiental-Especies nativas de México de flora y fauna silvestres-Categorías de riesgo y especificaciones para su inclusión, exclusión o cambio-Lista de especies en riesgo. Secretaría de Medio Ambiente y Recursos Naturales, México D.F.
DOF, Diario Oficial de los Estados Unidos Mexicanos, 2010b. Acuerdo mediante el cual se da a conocer la actualización de la Carta Nacional Pesquera. Secretaría de Medio Ambiente y Recursos Naturales, México D.F.
DOF, Diario Oficial de los Estados Unidos Mexicanos, 2011. Acuerdo por el que se da a conocer el resumen del Programa de Manejo de la Reserva de la Biosfera Isla Guadalupe. Secretaría de Medio Ambiente y Recursos Naturales, México D.F.

DOF, Diario Oficial de los Estados Unidos Mexicanos, 2012. Acuerdo por el que se modifica el Aviso por el que se da a conocer el establecimiento de épocas y zonas de veda para la pesca de diferentes especies de la fauna acuática en aguas de jurisdicción federal de los Estados Unidos Mexicanos, publicado el 16 de marzo de 1994 para establecer los periodos de veda de pulpo en el Sistema Arrecifal Veracruzano, jaiba en Sonora y Sinaloa, tiburones y rayas en el Océano Pacífico y tiburones en el Golfo de México. Secretaría de Agricultura, Ganadería, Desarrollo Rural, Pesca y Alimentación, México.

DOF, Diario Oficial de los Estados Unidos Mexicanos, 2013. ACUERDO Por el Que Se Da a Conocer el Resumen Del Programa de Manejo del Área Natural Protegida Con la Categoría de Reserva de la Biosfera la Zona Marina Conocida Como BAHÍA de Los Ángeles, Canales de Ballenas y de Salsipuedes. Secretaría de Medio Ambiente y Recursos Naturales, México D.F.

DOF, Diario Oficial de los Estados Unidos Mexicanos, 2014a. Acuerdo Por el Que se da a Conocer la Lista de Especies y Poblaciones Prioritarias Para la Conservación. Secretaría de Medio Ambiente y Recursos Naturales, México D.F.

DOF, Diario Oficial de los Estados Unidos Mexicanos, 2014b. Acuerdo Por el Que se Establece Veda Permanente Para la Pesca de Tiburón Blanco (Carcharodon carcharias) en Aguas de Jurisdicción Federal de Los Estados Unidos Mexicanos. Secretaría de Medio Ambiente y Recursos Naturales, México D.F.

DOF, Diario Oficial de los Estados Unidos Mexicanos, 2014c. Acuerdo Por el Que Se Modifica el Aviso Por el Que se da a CONOCER el establecimiento de Épocas y Zonas de Veda Para la Pesca de Diferentes Especies de la Fauna Acuática en Aguas de Jurisdicción Federal de los Estados Unidos Mexicanos, Publicado el 16 de marzo de 1994 Para Modificar el Periodo y Zonas de Veda de Tiburones en el Golfo de México y Mar Caribe. Secretaría de Medio Ambiente y Recursos Naturales, México D.F.

DOF, Diario Oficial de los Estados Unidos Mexicanos, 2014d. Reglamento de la Ley General del Equilibrio Ecológico y la Protección del Ambiente en materia de Áreas aturales Protegidas. Secretaría de Medio Ambiente y Recursos Naturales, México D.F.

DOF, Diario Oficial de los Estados Unidos Mexicanos, 2018a. Acuerdo Por el Que Se Da a Conocer la Actualización de la Carta Nacional Pesquera. Secretaría de Medio Ambiente y Recursos Naturales, México D.F.

DOF, Diario Oficial de los Estados Unidos Mexicanos, 2018b. ACUERDO Por el Que Se Da a Conocer el Resumen Del Programa de Manejo del Área Natural Protegida con la Categoría de Parque Nacional Revillagigedo. Secretaría de Medio Ambiente y Recursos Naturales, México D.F.

Domeier, M., Nasby-Lucas, N., 2008. Migration patterns of white sharks *Carcharodon carcharias* tagged at Guadalupe Island, Mexico, and identification of an eastern Pacific shared offshore foraging area. Mar. Ecol. Prog. Ser. 370, 221–237.

Dulvy, N.K., Fowler, S.L., Musick, J.A., Cavanagh, R.D., Kyne, P.M., Harrison, L.R., Carlson, J.K., Davidson, L.N.K., Fordham, S.V., Francis, M.P., Pollock, C.M., Simpfendorfer, C.A., Burgess, G.H., Carpenter, K.E., Compagno, L.J.V., Ebert, D.A., Gibson, C., Heupel, M.R., Livingstone, S.R., Sanciangco, J.C., Stevens, J.D., Valenti, S., White, W.T., 2014. Global extinction risk of sharks and rays: threats, hotspots, and conservation. eLife 3, e00590.

Durazo, R., Ramírez-Manguilar, A.M., Miranda, L.E., Soto-Mardones, L.A., 2007. Climatología de variables hidrográficas. In: Gaxiola-Castro, G., Durazo, R. (Eds.), Dinámica del Ecosistema Pelágico frente a Baja California, 1997–2007: Diez años de Investigaciones Mexicanas de la Corriente de California. Instituto Nacional de Ecología, México, pp. 25–57.

Ebert, D.A., White, W.T., Goldman, K.J., Compagno, L.J.V., Daly-Engel, T.S., Ward, R.D., 2010. Resurrection and redescription of *Squalus suckleyi* (Girard, 1854) from the North Pacifc, with comments on the *Squalus acanthias* subgroup (Squaliformes: Squalidae). Zootaxa 2612, 22–40.

Ebert, D.A., Fowler, S., Compagno, L.J.V., 2013. Sharks of the World: A Fully Illustrated Guide to the Sharks of the World. Wild Nature Press, Plymouth.

Ebert, D.A., Bigman, J.S., Lawson, J.M., 2017. Biodiversity, life history, and conservation of northeastern Pacific chondrichthyans. In: Larson, S.E., Lowry, D. (Eds.), Northeast Pacific Shark Biology. In: Research and Conservation Part A. Advances in Marine Biology, vol. 77. Academic Press, pp. 9–78.

Eckert, S.A., Stewart, B.S., 2001. Telemetry and satellite tracking of whale sharks, *Rhincodon typus*, in the Sea of Cortez, Mexico, and the North Pacific Ocean. Environ. Biol. Fish. 60, 299–308.

Ehemann, N.R., del Valle González-González, L., Chollet-Villalpando, J.G., De La Cruz-Agüero, J., 2018. Updated checklist of the extant Chondrichthyes within the exclusive economic zone of Mexico. ZooKeys 774, 17.

Espinosa, H., 2004. El Pacífico Mexicano. Cienc. UNAM Méx. 076, 14–21.

Espinosa-Pérez, H., 2014. Biodiversidad de peces en México. Rev. Mex. Biodiv. 85, 450–459.

Espinosa-Pérez, H., Castro-Aguirre, J.L., Huidobro-Campos, L., 2004. Catálogo sistemático de Tiburones (Elasmobranchii: Selachimorpha). Listados Faunísticos de México IX. Instituto de Biología UNAM, Ciudad de México, p. 135.

FAO, 1999. The International Plan of Action for the Conservation and Management of Sharks. Food and Agriculture Organization of the United Nations, Rome. http://www.fao.org/ipoa-sharks/en/.

Fernández-Méndez, J.I., González-Ania, L.V., Castillo-Géniz, J.L., 2016. Standardized catch rates for blue shark (Prionace glauca) in the 2006–2015 Mexican Pacific longline fishery based upon a shark scientific observer program. ISC Shark Working Group Workshop. ISC/16/SHARKWG-1/25.

Filonov, A.E., Tereshchenko, I.E., Monzón, C.O., González Ruelas, M.E., Godínez Domínguez, E., 2000. Variabilidad estacional de los campos de temperatura y salinidad en la zona costera de los estados de Jalisco y Colima, México. Cienc. Mar. 26 (2).

Furlong-Estrada, E., Tovar-Ávila, J., Ríos-Jara, E., 2014. Evaluación de riesgo ecológico de los tiburones capturados por la pesquería artesanal en la entrada del Golfo de California. Hidrobiológica 24 (2), 83–97.

Furlong-Estrada, E., Tovar-Ávila, J., Pérez-Jiménez, J.C., Ríos-Jara, E., 2015. Resiliencia de *Sphyrna lewini*, *Rhizoprionodon longurio* y *Carcharhinus falciformis* en la entrada del golfo de California después de tres décadas de explotación. Cienc. Mar. 41 (1), 49–63.

Furlong-Estrada, E., Galván-Magaña, F., Tovar-Ávila, J., 2017a. Use of the productivity and susceptibility analysis and a rapid management risk assessment to evaluate the vulnerability of sharks caught off the west coast of Baja California Sur, Mexico. Fish. Res. 194, 197–208.

Furlong-Estrada, E., Sosa-Nishizaki, O., García-Rodríguez, E., 2017b. Caracterización de la pesquería de tiburón azul (*Prionace glauca*) en el Pacífico Norte mexicano: un antecedente histórico. Cienc. Pesq. 25, 15–25.

Gallegos-García, A., Barberán-Falcón, J., 1998. Surgencia eólica. In: Tapia-García, M. (Ed.), El Golfo de Tehuantepec: El Ecosistema y Sus Recursos. In: 3, Universidad Autónoma Metropolitana-Iztapalapa, México, pp. 27–34.

Gallo, N.D., Levin, L.A., Beckwith, M., Barry, J.P., 2019. Home sweet suboxic home: remarkable hypoxia tolerance in two demersal fish species in the Gulf of California. Ecology 100 (3), 1–5.

Galvan-Magaña, F., Abitia-Cardenas, L.A., Rodríguez-Romero, J., Pérez-España, H., Chávez-Ramos, H., 1996. Lista sistemática de los peces de la isla Cerralvo, Baja California Sur, México. Cienc. Mar. 22 (3), 295–311.

Galván-Magaña, F., Hoyos-Padilla, E.M., Navarro-Serment, C.J., Márquez-Farías, F., 2010. Records of white shark, *Carcharodon carcharias*, in the Gulf of California, Mexico. Mar. Biodiv. Rec. 3 (e111), 1–6.

Galván-Tirado, C., Galván-Magaña, F., Ochoa-Báez, R.I., 2015. Reproductive biology of the silky shark *Carcharhinus falciformis* in the southern Mexican Pacific. J. Mar. Biol. Assoc. U. K. 95, 561–567.
García-Rodríguez, E., 2012. Análisis de la Pesquería de Tiburón Azul en la Costa Oeste de Baja California Con Inferencias en el Estado de la Población. Unpublished Master's thesis, Centro de Investigación Científica y de Educación Superior de Ensenada, México.
González-Ania, L.V., Fernández-Méndez, J.I., Castillo-Géniz, J.L., Hidalgo-García, L., Haro-Avalos, H., 2017. Update on Standardized Catch Rates for Mako Shark (Isurus oxyrinchus) in the 2006–2016 Mexican Pacific Longline Fishery Based Upon a Shark Scientific Observer Program. ISC Shark Working Group Workshop. ISC/17/SHA-RKWG-3-20.
González-Rodríguez, E., Trasviña-Castro, A., Ramos-Rodríguez, A., 2018. El bajo de espíritu santo; punto caliente de abundancia biológica afuera de bahía de la paz. CICIMAR *Oceánides* 33 (1), 13–24.
Graeber, R.C., 1974. Food intake patterns in captive juvenile lemon sharks, *Negaprion brevirostris*. Copeia 1974 (2), 554–556.
Guzmán, A.F., Meráz-Munguía, M.A., 2013. Vertebrados condrictios de la colección del Laboratorio de Ecología de la Escuela Nacional de Ciencias Biológicas, Instituto Politécnico Nacional, México. CICIMAR *Oceánides* 28 (2), 41–60.
Hastings, P.A., Findley, L.T., Van der Heiden, A.M., 2010. Fishes of the Gulf of California. In: Brusca, R. (Ed.), The Gulf of California Biodiversity and Conservation. University of Arizona Press, Tucson, pp. 96–118.
Heithaus, M.R., 2007. Nursery areas as essential shark habitats: a theoretical perspective. In: McCandless, C.T., Kohler, N.E., Pratt Jr., H.L. (Eds.), Shark Nursery Grounds of the Gulf of Mexico and the East Coast Waters of the United States. American Fisheries Society, Bethesda, Maryland, pp. 3–13.
Heupel, M.R., Simpfendorfer, C.A., 2009. Using acoustic monitoring to evaluate MPAs for shark nursery areas: the importance of long-term data. Mar. Technol. Soc. J. 39, 10–18.
Heupel, M.R., Carlson, J.K., Simpfendorfer, C.A., 2007. Shark nursery areas: concept, definition, characterization and assumptions. Mar. Ecol. Prog. Ser. 337, 287–297.
Heupel, M.R., Simpfendorfer, C.A., Fitzpatrick, R., 2010. Large-scale movement and reef fidelity of grey reef sharks. PLoS One 5 e9650.
Hillborn, R., 2017. Are MPAs effective? ICES J. Mar. Sci. 75, 1160–1162.
Hobday, A.J., Smith, A.D.M.C., Stobutzki, I., Bulman, C., Daley, R., Dambacher, J.M., Deng, R.A., Dowdney, J., Fuller, M., Furlani, D., Griffiths, S.P., Johnson, D., Kenyon, R., Knuckey, I.A., Ling, S.D., Pitcher, R., Sainsbury, K.J., Sporcic, M., Smith, T., Turnbull, C., Walker, T.I., Wayte, S.E., Webb, H., Williams, A., Wise, B.S., Zhou, S., 2011. Ecological risk assessment for the effects of fishing. Fish. Res. 108 (2-3), 372–384.
Holzwarth, S.R., DeMartini, E.E., Schroeder, R.E., Zgliczynski, B.J., Laughlin, J.L., 2006. Sharks and jacks in the Northwestern Hawaiian Islands from towed-diver surveys 2000–2003. Atoll Res. Bull. 543, 257–303.
Hubbs, C.L., Taylor, L.R., 1970. Data on life history and characters of *Galeus piperatus*, a dwarf shark of Golfo de California. Fiskeridirektorates Skrifter Serie Havundersokelser 15, 310–330.
IATTC, 2005. Resolution C-05-03: Resolution on the Conservation of Sharks Caught in Association With Fisheries in the Eastern Pacific Ocean. Inter-American Tropical Tuna Commission, La Jolla, USA. http://www.iattc.org/PDFFiles2/resolutions/C-05-03-Sharks.PDF.

IATTC, 2011. Resolution C-11-10: Resolution on the Conservation of Oceanic Whitetip Sharks Caught in Association with Fisheries in the Antigua Convention Area. Inter-American Tropical Tuna Commission, La Jolla, USA. https://www.iattc.org/PDFFiles2/Resolutions/C-11-10-Conservation-of-oceanic-whitetip-sharks.pdf.

IATTC, 2015. Resolution C-15-03: Collection and Analyses of Data on Fish-aggregating Devices. Inter-American Tropical Tuna Commission, La Jolla, USA. https://www.iattc.org/PDFFiles2/Resolutions/C-15-03-Amendment-C-13-04-FADs.pdf.

IATTC, 2016a. Active IATTC and AIDCP Resolutions and Recommendations. Inter-American Tropical Tuna Commission, La Jolla, USA. http://www.iattc.org/ResolutionsActiveENG.htm.

IATTC, 2016b. Resolution C-16-04: Amendment to Resolution c-05-03 on the Conservation of Sharks Caught in Association With Fisheries in the Eastern Pacific Ocean. Inter-American Tropical Tuna Commission, La Jolla, USA. https://www.iattc.org/PDFFiles2/Resolutions/C-16-04-Sharks-Amendment-C-05-03.pdf.

IATTC, 2016c. Resolution C-16-05: Resolution on management of shark species. Inter-American Tropical Tuna Commission, La Jolla, USA. https://www.iattc.org/PDFFiles2/Resolutions/C-16-05-Management-of-sharks.pdf.

ICCAT, 2010. Resolution 2010–08: Recommendation by ICCAT on Hammerhead Sharks (family Sphyrnidae) Caught in Association With Fisheries Managed by ICCAT. International Commission for the Conservation of Atlantic Tunas, Madrid, Spain.

ISC, 2017. Stock assessment and future projections of blue shark in the North Pacific Ocean through 2015. In: 17th meeting of the International Scientific Committee for Tuna and Tuna-Like Species in the North Pacific Ocean, Vancouver, Canada. 95 p.

ISC, 2018. Stock assessment of shortfin mako shark in the North Pacific Ocean through 2016. In: 18th meeting of the International Scientific Committee for Tuna and Tuna-Like Species in the North Pacific Ocean, Yeosu, Republic of Korea. 120 p.

IUCN, 2018. Applying IUCN's Global Conservation Standards to Marine Protected Areas (MPA). Delivering Effective Conservation Action Through MPAs, to Secure Ocean Health & Sustainable Development, Version 1.0. IUCN, Gland, Switzerland. 4 pp.

IUCN-SSG, 2019. Extinction Risk and Conservation of the World's Sharks and Rays. Analysis from the IUCN Shark Specialist Group. https://www.iucnssg.org/uploads/5/4/1/2/54120303/rl_assessment_results_190124.pdf.

Kato, S., 1965. White shark, *Carcharodon carcharias*, from the Gulf of California, with a list of the sharks in Mazatlan, Mexico. Copeia 1965 (3), 384.

Kinney, M.J., Simpfendorfer, C.A., 2009. Reassessing the value of nursery areas to shark conservation and management. Conser. Lett. 2, 53–60.

Klimley, A.P., 1981. Schooling of the scalloped hammerhead shark, *Sphyrna lewini*, in the Gulf of California. Fish. Bull. 79 (2), 356–360.

Klimley, A.P., 1987. The determinants of sexual segregation in the scalloped hammerhead, *Sphyrna lewini*. Environ. Biol. Fish. 18, 27–40.

Klimley, A.P., 2013. The Biology of Sharks and Rays. University of Chicago Press, Chicago. 512 pp.

Knip, D.M., Heupel, M.R., Simpfendorfer, C.A., 2012. Evaluating marine protected areas for the conservation of tropical coastal sharks. Biol. Conser. 148, 200–209.

Lara-Lara, J.R., Arenas Fuentes, V., Bazán Guzmán, C., Díaz Castañeda, V., Escobar Briones, E., García Abad, M.C., Gaxiola Castro, G., Robles Jarero, G., Sosa Ávalos, R., Soto González, L.A., Tapia, 2008. Los ecosistemas marinos. In: Comisión Nacional para el Conocimiento y Uso de la Biodiversidad (CONABIO), (Ed.), Capital Natural de México. In: vol. I. Conocimiento actual de la biodiversidad. CONABIO, México, pp. 135–159.

Lluch-Cota, S.E., Aragón-Noriega, E.A., Arreguín-Sánchez, F., Aurioles-Gamboa, D., Bautista-Romero, J.J., Brusca, R.C., Cervantes-Duarte, R., Cortés-Altamirano, R., Del Monte-Luna, P., Esquivel-Herrera, A., Fernández, G., Hendrickx, M.E., Hernández Vázquez, S., Herrera Cervantes, H., Kahru, M., Lavín, M., Lluch-Belda, D., Lluch-Cota, D.B., López-Martínez, J., Marinote, S.G., Nevárez-Martínez, M.O., Ortega-García, S., Palacios Castro, E., Parés-Sierra, A., Ponce-Díaz, G., Ramírez Rodríguez, M., Salinas Zavala, C.A., Schwartzlose, R.A., Sierra-Beltrán, A.P., 2007. The Gulf of California: review of ecosystem status and sustainability challenges. Progr. Oceanogr. 73, 1–26.

Long, D.J., Seigel, J.A., 1997. A crocodile shark, Pseudocarcharias kamoharai (Selachii: Lamnidae) from pelagic waters off Baja California, México. Oceánides 12 (1), 61–63.

Long, D.J., McCosker, J.E., Blum, S., Klapfer, A., 2011. Tropical Eastern Pacific records of the prickly shark, *Echinorhinus cookei* (Chondrichthyes: Echinorhinidae). Pac. Sci. 65 (4), 433–441.

Márquez-Farías, J.F., 2000. Tiburones del Golfo de California. In: Cisneros-Mata, M.A., Beléndez-Moreno, L.F., Zárate-Becerra, E., Gaspar-Dillanes, M.T., López-González, L.C., Saucedo-Ruíz, C., Tovar-Áavila, J. (Eds.), Sustentabilidad y Pesca Responsible en México: Evaluación y Manejo 1999–2000. Instituto Nacional de Pesca, SEMARNAP, México, pp. 238–257.

Mejía-Mercado, B.E., Balart-Páez, E.F., Sosa-Nishizaki, O., Hinojosa-Corona, A., 2013. Catálogo, Foto–Identifcación de la Fauna Íctica de Profundidad (350–3800 m) del Golfo de California. CICESE, Ensenada.

Mondragón-Sánchez, L.F., 2011. Análisis de la Composición de Tallas y Edades de los Tiburones Capturados por la Pesquería Artesanal de Nayarit (2007–2010). Unpublished undergraduate thesis, Universidad del Mar.

Mondragón-Sánchez, L.F., 2016. Demografía del Tiburón Sedoso, *Carcharhinus falciformis* (Müller y Henle, 1839), en el PACÍFICO Central Mexicano. Tesis de maestría, UNAM. 68 p.

Mondragón-Sánchez, L.F., Tovar-Ávila, J., Castillo-Géniz, J.L., Fernández-Méndez, J.I., y González-Ania, L.V., 2018. Preliminary Demographic Analysis of Shortfin Mako Shark (Isurus oxyrinchus) in the Mexican Pacific Ocean. ISC Shark Working Group Workshop. ISC/18/SHARKWG-2/04.

Morales-Pacheco, O., Hernández-Corona, J.L., Lizárraga-Rodríguez, A.A., Alatorre-Alba, A.J., 2006. Golfo de tehuantepec. In: Castillo-Géniz, J.L., Tovar Ávila, J. (Eds.), Tiburones Mexicanos de Importancia Pesquera en la CITES. Instituto Nacional de Pesca, México, pp. 53–55.

Nelson, J.S., Grande, T.C., Wilson, M.V., 2016. Fishes of the World, fifth ed. John Wiley & Sons Hoboken, NJ.

Oñate-González, E.C., Rocha-Olivares, A., Saavedra-Sotelo, N.C., Sosa-Nishizaki, O., 2015. Mitochondrial genetic structure and matrilineal origin of white sharks, *Carcharodon carcharias*, in the Northeastern Pacific: implications for their conservation. J. Hered. 106, 347–354.

Oñate-González, E.C., Sosa-Nishizaki, O., Herzka, S.Z., Lowe, C.G., Lyons, K., Santana-Morales, O., Sepulveda, C., Guerrero-Ávila, C., García-Rodríguez, E., O'Sullivan, J.B., 2017. Importance of Bahia Sebastian Vizcaino as a nursery area for white sharks (*Carcharodon carcharias*) in the Northeastern Pacific: A fishery dependent analysis. Fish. Res. 188, 125–137.

Palacios-Salgado, D.S., Burnes-Romo, L.A., Tavera, J.J., Ramírez-Valdez, A., 2012. Endemic fishes of the Cortez biogeographic province (Eastern Pacific Ocean). Acta Ichthyol. Piscat. 42, 153–164.

Patrick, W.S., Spencer, P., Link, J., Cope, J., Field, J., Kobayashi, D., Lawson, P., Gedamke, T., Cortés, E., Ormseth, O., Bigelow, K., Overholtz, W., 2010. Using productivity and susceptibility indices to assess the vulnerability of United States fish stocks to overfishing. Fish. Bull. 108, 305–322.

Pérez-Jiménez, J.C., 2014. Historical records reveal potential extirpation of four hammerhead sharks (*Sphyrna* spp.) in Mexican Pacific waters. Rev. Fish Biol. Fish. 24, 671–683.

Pérez-Jiménez, J.C., Sosa-Nishizaki, O., 2010. Determining reproductive parameters for population assessments of two smoothhounds (*Mustelus californicus* and *Mustelus lunulatus*) from the northern Gulf of California, Mexico. Bull. Mar. Sci. 86, 3–13.

Pérez-Jiménez, J.C., Sosa-Nishizaki, O., Furlong-Estrada, E., Corro-Espinosa, D., Venegas-Herrera, A., Barragán-Cuencas, O.V., 2005. Artisanal shark fishery at Tres Marias Islands and Isabel Island in the Central Mexican Pacific. J. Northw. Atl. Fish. Sci. 35, 333–343.

Ramírez-Amaro, S.R., Cartamil, D., Galván-Magaña, F., González-Barba, G., Graham, J.B., Carrera-Fernandez, M., Escobar-Sanchez, O., Sosa-Nishizaki, O., Rochin-Alamillo, A., 2013. The artisanal elasmobranch fisheries in the Pacific coast of Baja California Sur, Mexico, management implications. Sci. Mar. 77 (3), 473–487.

Ramírez-Macias, D., Vázquez Haikin, A., Vázquez Juárez, R., 2012. Whale shark *Rhincodon typus* populations along the west coast of the Gulf of California and implications for management. Endanger. Species Res. 18, 115–128.

Ramírez-Macías, D., Queiroz, N., Pierce, S.J., Humphries, N.E., Sims, D.W., Brunnschweiler, J.M., 2017. Oceanic adults, coastal juveniles: tracking the habitat use of whale sharks off the Pacific coast of Mexico. PeerJ 5, e3271.

Reyes-Bonilla, H., Ayala-Bocos, A., González-Romero, S., Sánchez-Alcántara, I., Walther-Mendoza, M., Bedolla-Guzman, Y.R., Ramírez-Valdez, A., Calderón Aguilera, L.E., Olivares-Banuelos, N.C., 2010. Checklist and biogeography of fishes from Guadalupe Island, Western México. CalCOFI Rep. 51, 195–209.

Rodríguez-Madrigal, J.A., 2018. Estimación de Crecimiento Del Tiburón Mako (*Isurus oxyrinchus*) en el Pacífico Mexicano Mediante Inferencia Multimodal y Distintos Métodos Para Estimar la Edad. Unpublished Master's thesis, Posgrado de Ciencias del Mar y Limnología, UNAM, México.

Rodríguez-Madrigal, J.A., Semba, Y., Tovar-Ávila, J., 2017a. Standardization of Mako Shark Aging Through Different Vertebrae Enhancement Methods and Comparison of Growth Estimations From Eastern and Western North Pacific Ocean. ISC Shark Working Group Workshop. ISC/17/SHARKWG-3/16.

Rodríguez-Madrigal, J.A., Tovar-Ávila, J., Castillo-Géniz, J.L., Godínez-Padilla, C.J., Galván-Magaña, F., Márquez-Farías, J.F., Corro-espinosa, D., 2017b. Growth Estimation Update of Shortfin Mako Shark in the Mexican Pacific Ocean, Through Multi-Model Approach and Different Methods for Age Determination. ISC Shark Working Group Workshop. ISC/17/SHARKWG-1/15.

Rodríguez-Romero, J., Palacios-Salgado, D.S., López-Martínez, J., Hernández-Vázquez, S., Ponce-Díaz, G., 2008. Composición taxonómica y relaciones zoogeográfcas de los peces demersales de la costa occidental de Baja California Sur, México. Rev. Biol. Trop. 56 (4), 1765–1783.

Rosende-Pereiro, A., Corgos, A., 2018. Pilot acoustic tracking study on young of the year scalloped hammerhead sharks, *Sphyrna lewini*, within a coastal nursery area in Jalisco, Mexico. Lat. Am. J. Aquat. Res. 46 (4), 645–659.

Ruiz-Campos, G., Castro-Aguirre, J.L., Balart, E.F., Campos-Dávila, L., Vélez-Marín, R., 2010. New specimens and records of chondrichthyan fishes (Vertebrata: Chondrichthyes) off the Mexican Pacific coast. Rev. Mexic. Biodiv. 81 (2), 363–371.

Saldaña-Ruiz, L.E., Sosa-Nishizaki, O., Cartamil, D., 2017. Historical reconstruction of Gulf of California shark fishery landings and species composition, 1939–2014, in a data-poor fishery context. Fish. Res. 195, 116–129.

Salomón-Aguilar, C., Villavicencio-Garayzar, C.J., Reyes-Bonilla, H., 2009. Shark breeding grounds and seasons in the Gulf of California: fishery management and conservation strategy. Cienc. Mar. 35, 369–388.

Sandoval-Castillo, J., Ramírez-González, J., Villavicencio-Garayzar, C., 2008. First record of basking shark (*Cetorhinus maximus*) in Mexico? Mar. Biodiv. Rec. 1 (e19), 1–2.

Santana-Hernández, H., Flores-Valdez, J., 2014. Pelágicos Mayores Obtenidos Por la Flota Palangrera de Mediana Altura del Puerto de Manzanillo, Colima. Instituto Nacional de Pesca, México.

Santana-Hernández, H., Espino-Barr, E., Valdez-Flores, J.J., 2013. Distribución y abundancia relativa del tiburón oceánico de aletas blancas *Carcharhinus longimanus*, capturado por barcos palangreros en el Pacífico central mexicano. Cienc. Pesq. 21 (1), 27–39.

Santana-Hernández, H., Tovar-Ávila, J., Valdéz-Flores, J.J., 2014. Estimation of the total, fork and precaudal lengths for the silky shark, *Carcharhinus falciformis* (Carcharhiniformes: Carcharhinidae), from the interdorsal length. Hidrobiológica 24 (2), 159–162.

Santos del Prado, K., Peters, E., 2005. La Isla Guadalupe. Restauración y Conservación, SEMARNAT/INE/CICESE/GECI/SEMAR, México.

Secretaría de Agricultura, Ganadería, Desarrollo Rural, Pesca y Alimentación (SAGARPA), 2016. Anuario Estadístico de Acuacultura y Pesca 2017. Comisión Nacional de Acuacultura y Pesca, México.

Smith, W.D., Bizzarro, J.J., Cailliet, G.M., 2009. The artisanal elasmobranch fishery on the east coast of Baja California, Mexico: characteristics and management considerations. Cienc. Mar. 35 (2), 209–236.

Soriano-Velásquez, S.R, Acal-Sánchez, D., Castillo-Géniz, J.L., Ramírez-Santiago, C.E., 2006. Tiburón del Golfo de Tehuantepec. In: Arreguín, S.F., Meléndez, M.J., Gómez-Humarán, M.I., Solana, S.R., Rangel, D.C. (Eds.), Sustentabilidad y Pesca Responsable de México, Evaluación y Manejo 1999–2000. Instituto Nacional de Pesca, México, pp. 325–364.

Sosa-Nishizaki, O., Castillo-Géniz, J.L., 2016. Blue Shark Catches Estimations for the Mexican Pacific (1976–2014). ISC Shark Working Group Workshop. ISC/16/SHARKWG-1/24.

Sosa-Nishizaki, O., Márquez-Farías, J.F., Villavicencio-Garayzar, C., 2008. Case study: Pelagic shark Fisheries along the west coast of Mexico. In: Camhi, M.D., Pikitch, E.K., Babcock, E.A. (Eds.), Sharks of the Open Ocean, Biology, Fisheries & Consevation. In: Fish and Aquatic Resources Series, 13, Blackwell, pp. 275–282.

Sosa-Nishizaki, O., Saldaña-Ruiz, L.E., Corro-Espinosa, D., Tovar-Ávila, J., Castillo-Géniz, J.L., Santana-Hernández, H., y Márquez-Farías, J.F., 2014. Estimations of the Shortfin Mako Shark (*Isurus oxyrinchus*) catches by Mexican Pacific Fisheries (1976–2013). ISC Shark Working Group Workshop. ISC/14/SHARKWG-3/17.

Sosa-Nishizaki, O., Saldaña-Ruiz, L.E., Corro-Espinosa, D., Tovar-Ávila, J., Castillo-Géniz, J.L., Santana-Hernández, H., Márquez-Farías, J.F., 2017. Estimations of the Shortfin Mako Shark (Isurus oxyrinchus) Catches by Mexican Pacific fisheries. ISC Shark Working Group Workshop. An update (1976–2016). ISC/17/SHARKWG-3/19.

IUCN Standards and Petitions Subcommittee, 2017. Guidelines for Using the IUCN Red List Categories and Criteria. Version 13. Prepared by the Standards and Petitions Subcommittee. IUCN-SSC. Downloaded from http://www.iucnredlist.org/documents/RedListGuidelines.pdf.

Stevenson, D.E., Orr, J.W., Hoff, G.R., McEachran, J.D., 2007. Field Guide to Sharks, Skates, and Ratfsh of Alaska. Alaska Sea Grant College Program.

Teo, S.L.H., García-Rodríguez, E., Sosa-Nishizaki, O., 2018. Status of Common Thresher Shark, Alopias Vulpinus, Along the West Coast of North America: Updated Stock Assessment Based on Alternative Life History. NOAA Technical Memorandum NMFS (NOAA-TM-NMFS-SWFSC-595).

Torres Huerta, A.M., Villavicencio Garayzar, C., Corro Espinosa, D., 2008. Biología reproductiva de la cornuda común *Sphyrna lewini* Griffith & Smith (Sphyrnidae) en el Golfo de California. Hidrobiológica 18 (3), 227–238.

Tovar Ávila, J., Furlong Estrada, E., Castillo Géniz, J.L., 2016. Evaluación de riesgo ecológico por efectos de las pesquerías de tiburón mexicanas para las especies incluidas en el Apéndice II de la CITES. In: Castillo-Géniz, J.L., Tovar-Ávila, J. (Eds.), Tiburones Mexicanos de Importancia Pesquera en la CITES. INAPESCA, Mexico, pp. 17–23.

Tovar-Ávila, J., González-Ania, L., Liedo-Galindo, A., Márquez-Farías, F., 2011. Outline of New Available Catch and Effort Data of Pelagic Sharks Caught by the Mexican Shark Longline Fishery in the North Pacific. ISC Shark Working Group Workshop. ISC/11/SHARKWG-1/08.

Trasviña, A., Barton, E.D., Brown, J., Vélez, H.S., Kosro, M., Smith, R.L., 1995. Offshore Wind Forcing in the Gulf of Tehuantepec, Mexico: the asymmetric circulation. J. Geophys. Res. OCEANS 100 (C10), 20649–20,663.

Trasviña, A., Lluch-Cota, D., Filonov, A.E., Gallegos, A., 1999. Oceanografía y el Niño. In: Magaña, R.V. (Ed.), Los Impactos de El Niño en México. UNAM, México, pp. 69–101.

Trejo-Ramírez, A., 2017. Caracterización de la Bahía de La Paz, B. C. S., México, como una posible área de crianza del tiburón bironche, Rhizoprionodon longurio. Unpublished Master's thesis, CICMAR, La Paz, B.C.S., México. 147 pp.

Vélez-Marín, R., Márquez-Farías, J.F., 2009. Distribution and size of the shortfin mako (*Isurus oxyrinchus*) in the Mexican Pacific Ocean. Pan Am. J. Aquat. Sci. 4 (4), 490–499.

Villavicencio-Garayzar, C.J., 1996. Aspectos poblacionales del angelito, *Squatina californica* (Ayres), en Baja California, México. Rev. Invest. Cien. Ser. Cienc. Mar. UABCS 7 (1–2), 15–21.

Vincent, A.C.J., Sadovy de Mitcheson, Y., Fowler, S.L., Lieberman, S., 2014. The role of CITES in the conservation of marine fishes subject to international trade. Fish Fish. 15, 563–592.

Vögler, R., Beier, E., Ortega-García, S., Santana-Hernández, H., Valdez-Flores, J.J., 2012. Ecological patterns, distribution and population structure of *Prionace glauca* (Chondrichthyes: Carcharhinidae) in the tropical-subtropical transition zone of the north-eastern Pacific. Mar. Environ. Res. 73, 37–52.

Walker, T.I., 2005. Management measurements. In: Musick, J.A., Bonfil, R. (Eds.), Management Techniques for Elasmobranch Fisheries. Food and Agriculture Organization of the United Nations. Roma, Italia, pp. 216–242.

Weigmann, S., 2016. Annotated checklist of the living sharks, batoids, and chimaeras (Chondrichthyes) of the world, with a focus on biogeographical diversity. J. Fish Biol. 88, 837–1037.

Weng, K.C., O'Sullivan, J.B., Lowe, C.G., Winkler, C.E., Dewar, H., Block, B.A., 2007. Movements, behavior and habitat preferences of juvenile white sharks *Carcharodon carcharias* in the Eastern Pacific. Mar. Ecol. Prog. Ser. 338, 211–224.

Weng, K.C., O'Sullivan, J.B., Lowe, C.G., Winkler, C.E., Blasius, M.E., Loke-Smith, K.A., Sippel, T.J., Ezcurra, J.M., 2012. Back to the wild. Release of juvenile white sharks from the Monterey Bay Aquarium. In: Domeier, M.L. (Ed.), Global Perspectives on the Biology and Life History of White Shark. CRC Press, New York, USA, pp. 419–446.

White, T.D., Ferretti, F., Kroodsma, D.A., Hazen, E.L., Carlisle, A.B., Scales, K.L., Bograd, S.J., Block, B.A., 2019. Predicted hotspots of overlap between highly migratory fishes and industrial fishing fleets in the Northeast Pacific. Sci. Adv. 5 (3) eaau3761.

Wilkinson, T., Wiken, E., Bezaury Creel, J., Hourigan, T., Agardy, T., Herrmann, H., Janishevski, L., Madden, C., Morgan, L., Padilla, M., 2009. Ecorregiones marinas de América del Norte, Comisión para la Cooperación Ambiental. Montreal, Canada.

## Further reading

Castillo-Géniz, J.L., Tovar-Avila, J., 2016. Tiburones mexicanos de importancia pesquera en la CITES. Instituto Nacional de la Pesca, México.

Secretaría de Agricultura, Ganadería, Desarrollo Rural, Pesca y Alimentación (SAGARPA), 2008. Anuario Estadístico de Acuacultura y Pesca 2009. Comisión Nacional de Acuacultura y Pesca, México.

CHAPTER THREE

# Shark ecology, the role of the apex predator and current conservation status


Felipe Galván-Magaña[a,*], José Leonardo Castillo-Geniz[b], Mauricio Hoyos-Padilla[c], James Ketchum[c], A. Peter Klimley[d], Sergio Ramírez-Amaro[e,f], Yassir Eden Torres-Rojas[g], Javier Tovar-Ávila[h]

[a]Instituto Politécnico Nacional, Centro Interdisciplinario de Ciencias Marinas, La Paz, Mexico
[b]Instituto Nacional de Pesca y Acuacultura, National Fisheries and Aquaculture Institute, Centro Regional de Investigación Pesquera Ensenada, La Paz, Mexico
[c]Pelagios-Kakunjá, La Paz, Mexico
[d]Biotelemetry Consultants and Contractors, Petaluma, CA, United States
[e]Instituto Español de Oceanografía, Centre Oceanogràfic de les Balears, Palma, Spain
[f]Laboratori de Genètica, Universitat de les Illes Balears, Palma, Spain
[g]Instituto de Ecología, Pesquerías y Oceanografía del Golfo de México, Universidad Autónoma de Campeche (EPOMEX-UAC), Campeche, Mexico
[h]Instituto Nacional de Pesca (INAPESCA), Centro Regional de Investigacion Pesquera (CRIP), La Cruz de Huanacaxtle, Nayarit, Mexico
*Corresponding author: e-mail address: galvan.felipe@gmail.com


## Contents



## Abstract


Feeding studies, since traditional stomach content analysis to stable isotopes analyses, provides insights into the trophic relationship among the apex predators and the ecosystems they inhabit. The Pacific Coast of Mexico (PCM) is inhabited by 62 known species (or 12%) of living sharks, which belong to 21 families and 34 genera. We divide the Pacific Coast of Mexico (PCM) into four regions for consideration: (1) the western coast



*Advances in Marine Biology*, Volume 83
ISSN 0065-2881
https://doi.org/10.1016/bs.amb.2019.08.005

of Baja California (WcBJ), (2) the Gulf of California (GC), (3) the Central Pacific Mexican (CPM), and (4) the Gulf of Tehuantepec (GT). Biodiversity is highest in the GC, with 48 shark species, followed by the WcBJ with 44 species, then the CPM with 28 species and the GT with 26 species. Few large species (>2 m in total length) function as top predators in any region, with a greater number of smaller shark species (<1.5 m total length). Information about the trophic ecology of different shark species is included to know the ecological role and position of each shark species within a food web to understand the dynamics of marine communities and the impact that each species has on trophic net, which is critical to effective resource conservation and responsible exploitation. The different shark species predate mainly on coastal or oceanic waters. The coastal sharks feed mainly on crustaceans and small fishes; whereas the oceanic species predate mainly on squids and fishes from mesopelagic to epipelagic habits. Also is included a summary of the IUCN Red List category assigned to all shark species from the Mexican Pacific. Thirty-one percent (19 species) of sharks in the Mexican Pacific are considered as threatened (Critically Endangered, Endangered or Vulnerable). Of these, 4.9% (3 species) are Endangered and 26.2% (15 species) are Vulnerable. In addition, since 2012 the fishing of shark and rays has been closed between 1 May and 31 July in the Mexican Pacific as a conservative management measure.

## 1. Introduction

The Northeast Pacific Ocean (NEP) is an extensive marine region encompassing the following large marine ecosystems: Eastern Bering Sea, Gulf of Alaska, California Current System, and Pacific Coast of Mexico (including the Gulf of California) (Fig. 1). The NEP includes sub-polar, temperate, subtropical and tropical realms. Its oceanography is varied and is influenced either by the cool North Pacific Drift or the warm Equatorial Counter Current. The Pacific Coast of Mexico (PCM) is the largest area within the NEP, with nearly 3117 km of coastline. This area is influenced by the Equatorial Counter Current that flows northward along the coast of Central America and brings tropical water up the coast of Mexico to the mouth of the Golf of California and the southern tip of Baja California Sur. A portion of this warm water continues northward along the western coast of the Baja Peninsula as far as Southern California, forming a gyre in the Southern California Bight (Zaytsev et al., 2003). The climate off the Pacific Coast of Mexico ranges from temperate to subtropical, becoming tropical in the South of the region where marine ecosystem diversity is generally greater. In the southern part of this area, many tropical habitats such as coral reefs, sandy and rocky shores, submarine canyons, estuaries, mangroves and coastal lagoons provide habitat for a wealth of marine species. A number of 891 species of fish have been reported in the Gulf of California (Brusca et al., 2005).

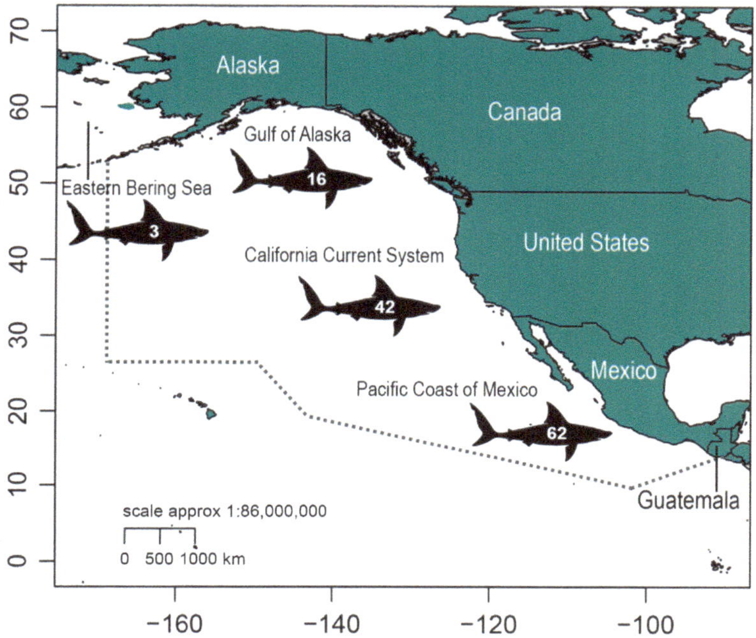


**Fig. 1** Map of the Northeast Pacific Ocean (NEP). Sharks icons show the number of shark species for each of four marine ecosystems: Eastern Bering Sea, Gulf of Alaska, California Current System, and Pacific Coast of Mexico. Dotted lines indicate the edge of the Exclusive Economic Zone.

The economically important activities associated with the ocean along the Pacific Coast of Mexico are varied. The northern region is characterized by large-scale, profitable fisheries on species such as sardines, anchovies, and tuna, and coastal tourism has increased in recent years. The Gulf of California is the home of important fishing ports such as Mazatlan, Guaymas, Puerto Peñasco and Topolobampo as well as tourist destinations such as Los Cabos, La Paz, and Loreto. The coastal zone of the Baja peninsula has been impacted by mineral extraction at locations such as Santa Rosalia and La Paz. The economic activity along the southwestern coast of Pacific Mexico is confined to tourism with resorts in the cities of Manzanillo, Zihuatanejo, Ixtapa, Acapulco, Huatulco and Puerto Angel (Rivera-Arriaga and Villalobos, 2001).

For the purposes of this chapter, we divide the PCM into the four following regions: western coast of Baja California (WcBJ); Gulf of California (GC); Central Pacific Mexican (CPM), and Gulf of Tehuantepec (GT) (Fig. 2). The WcBJ is a region characterized by high productivity and diversity of marine resources, which are influenced by the California Current

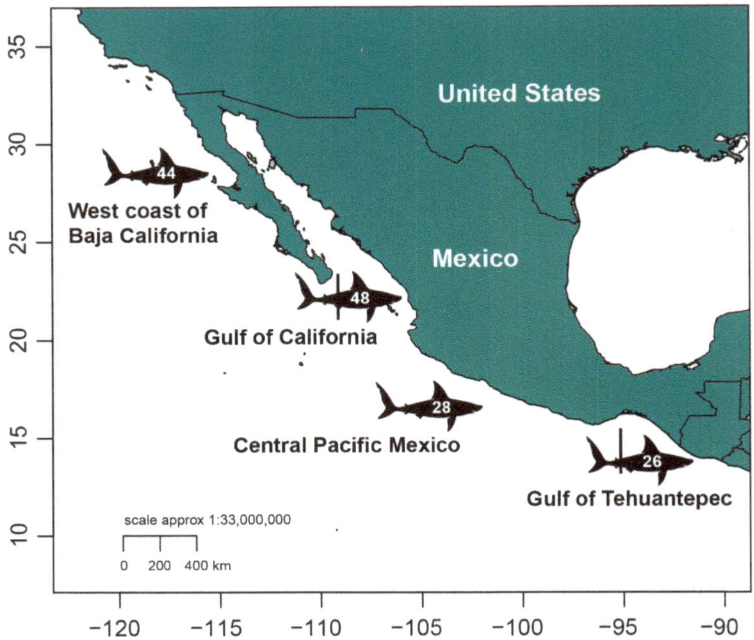


**Fig. 2** Map of the Pacific Coast of Mexico (PCM). Shark icons show the number of shark species for four marine areas: Western coast of Baja California, Gulf of California, Central Pacific Mexico, and Gulf of Tehuantepec.

System that flows southward and generates costal upwelling along this region (Young, 2001; Zaytsev et al., 2003). The GC is a large, semi-enclosed sea extending 1100 km in length and 100–200 km in width. It is home to >900 islands are notable as a region of special biological significance due to its great habitat diversity (Brusca et al., 2005). The CPM is an oceanographic transition zone that combines fauna from the northern and southern regions of the Mexican Pacific. The topography of this region is complex due to the presence of extensive bays, coastal lagoons, estuaries, mangroves and coral reefs (Arriaga-Cabrera et al., 1998). The GT is large body water on the Pacific coast of the Isthmus of Tehuantepec in southeastern Mexico. A strong wind, the 'Tehuanos', blows periodically out over the waters of the GT, causing upwelling of nutrient-rich water that supports abundant sea life (Molina-Cruz and Martinez-López, 1994).

## 2. Sharks of the Pacific Coast of Mexico

Globally, there are ca. 524 known species of sharks (Ebert et al., 2017). The PCM is inhabited by 62 known species (or 12%) of living sharks, which belong to 21 families and 34 genera (Ehemann et al., 2018). Carcharhinidae is the most taxonomically diverse family with 8 genera and 19 species (Table 1; Fig. 3). Of these species, 12 belong to the genus *Carcharhinus*. Triakidae, the second most diverse family, is composed of three genera and seven species. Among the species in the PCM are iconic species from other families such as the Whale Shark *Rhincodon typus*, White Shark *Carcharodon carcharias*, and Basking Shark *Cetorhinus maximus*. There are only two known endemic species in the region: the Whitemargin Smoothhound *Mustelus albipinnis* and Peppered Catshark *Galeus piperatus*. The PCM has greater diversity of sharks compared with other major ocean regions across the NEP, such as the California Current System with 42 species of sharks, the Gulf of Alaska with 16 species, and the Eastern Bering Sea with 3 species (Ebert, 2003; Ebert et al., 2017) (Fig. 1).

The diversity of sharks within the PCM is not homogeneous. It is greatest in the GC with 48 species and the WcBJ with 44 species, followed by the CPM with 28 species and the GT with 26 species (Anislando-Tolentino et al., 2008; Bizzarro et al., 2009a,b,c; Cartamil et al., 2011; Castillo-Geniz et al., 2002; Ehemann et al., 2018; Ramirez-Amaro et al., 2013; Saldaña-Ruiz et al., 2017; Smith et al., 2009a; Soriano-Velásquez et al., 2006) (Fig. 2). Only 14 species of shark have been recorded in all 4 regions of the PCM (Table 1). This pattern of diversity is caused by ecological and environmental factors such as habitat diversity, sea temperature, salinity, and seafloor bathymetry. Climate change is likely to increase variation in many of these factors and affect shark populations and local species composition and, therefore, the functioning of the ecosystem at large (Harley, 2011; Pistevos et al., 2015).

The species composition in the PCM is dominated by coastal species (65%), whereas pelagic/oceanic (19%) and deep-sea species (16%) are less diverse (Table 1). Generally, species within the genus *Carcharhinus* occupy the coastal habitat, with the pelagic habitat occupied by members of the Alopiidae and Lamnidae. However, the habitats visited by these species vary temporally, spatially, and ontogenetically, and it is important to document and understand variation in this habitat dynamism for management and conservation of shark species (Grubbs, 2010).

Weighmann (2016) did not include the PCM as a 'hotspot' of chondrichthyan biodiversity. However, with the number of ray, skate, and chimaera species already reported from the region, Mexico's Eastern Pacific may have one of the highest diversities of chondrichthyan fishes in the world. Coastal, neritic, and demersal shark species are the most well-studied group because they are the species most often targeted by artisanal

**Table 1** List of sharks documented in the West coast of Baja California (WcBJ), Gulf of California (GC), Central Pacific Mexico (CPM), and Gulf of Tehuantepec (GT).

| Species | Habitat | Family | WcBJ | GC | CPM | GT |
|---|---|---|---|---|---|---|
| *Alopias pelagicus* | Pelagic/ Oceanic | Alopiidae | X | X | X | X |
| *Alopias superciliosus* | Pelagic/ Oceanic | Alopiidae | | X | X | X |
| *Alopias vulpinus* | Pelagic/ Oceanic | Alopiidae | X | X | X | X |
| *Apristurus brunneus* | Deep sea | Scyliorhinidae | | X | | |
| *Apristurus kampae* | Deep sea | Scyliorhinidae | | X | | |
| *Apristurus nasutus* | Deep sea | Scyliorhinidae | | X | | |
| *Carcharhinus albimarginatus* | Coastal | Carcharhinidae | | | X | |
| *Carcharhinus altimus* | Coastal | Carcharhinidae | X | X | X | X |
| *Carcharhinus brachyurus* | Coastal | Carcharhinidae | X | X | X | X |
| *Carcharhinus cerdale* | Coastal | Carcharhinidae | | X | | |
| *Carcharhinus falciformis* | Coastal | Carcharhinidae | X | X | X | X |
| *Carcharhinus galapagensis* | Coastal | Carcharhinidae | | X | X | |
| *Carcharhinus leucas* | Coastal | Carcharhinidae | X | X | X | X |
| *Carcharhinus limbatus* | Coastal | Carcharhinidae | X | X | X | X |
| *Carcharhinus longimanus* | Pelagic/ Oceanic | Carcharhinidae | X | X | X | X |
| *Carcharhinus obscurus* | Coastal | Carcharhinidae | X | X | X | X |
| *Carcharhinus porosus* | Coastal | Carcharhinidae | | X | | X |
| *Carcharhinus plumbeus* | Coastal | Carcharhinidae | | | X | |
| *Carcharodon carcharias* | Pelagic/ Oceanic | Lamnidae | X | X | | |
| *Cephaloscyllium ventriosum* | Coastal | Scyliorhinidae | X | X | | |
| *Cephalurus cephalus* | Deep sea | Pentanchidae | | X | X | |
| *Centroscyllium nigrum* | Deep sea | Squalidae | X | | | |

**Table 1** List of sharks documented in the West coast of Baja California (WcBJ), Gulf of California (GC), Central Pacific Mexico (CPM), and Gulf of Tehuantepec (GT).—cont'd

| Species | Habitat | Family | WcBJ | GC | CPM | GT |
|---|---|---|---|---|---|---|
| *Cetorhinus maximus* | Pelagic/ Oceanic | Cetorhinidae | X | | | |
| *Chlamydoselachus anguineus* | Deep sea | Chlamydoselachidae | X | | X | |
| *Echinorhinus cookei* | Coastal | Echinorhinidae | X | X | | |
| *Galeocerdo cuvier* | Coastal | Carcharhinidae | X | X | X | X |
| *Galeorhinus galeus* | Coastal | Triakidae | X | X | | |
| *Galeus piperatus* | Deep sea | Scyliorhinidae | | X | | |
| *Ginglymostoma cirratum* | Coastal | Ginglymostomatidae | | X | X | X |
| *Ginglymostoma unami* | Coastal | Ginglymostomatidae | | X | | |
| *Heterodontus francisci* | Coastal | Heterodontidae | X | X | | |
| *Heterodontus mexicanus* | Coastal | Heterodontidae | X | X | | |
| *Hexanchus griseus* | Deep sea | Hexanchidae | X | X | | |
| *Isistius brasiliensis* | Deep sea | Dalatiidae | X | | | |
| *Isurus oxyrinchus* | Pelagic/ Oceanic | Lamnidae | X | X | X | X |
| *Isurus paucus* | Pelagic/ Oceanic | Lamnidae | X | | | |
| *Lamna ditropis* | Pelagic/ Oceanic | Lamnidae | X | | | |
| *Megachasma pelagios* | Pelagic/ Oceanic | Megachasmidae | X | | | |
| *Mustelus albipinnis* | Coastal | Triakidae | X | X | | |
| *Mustelus californicus* | Coastal | Triakidae | X | | | |
| *Mustelus dorsalis* | Coastal | Triakidae | X | | | |
| *Mustelus henlei* | Coastal | Triakidae | X | X | | |
| *Mustelus lunulatus* | Coastal | Triakidae | X | X | | X |
| *Nasolamia velox* | Coastal | Carcharhinidae | X | X | X | X |
| *Negaprion brevirostris* | Coastal | Carcharhinidae | | X | X | X |

*Continued*

**Table 1** List of sharks documented in the West coast of Baja California (WcBJ), Gulf of California (GC), Central Pacific Mexico (CPM), and Gulf of Tehuantepec (GT).—cont'd

| Species | Habitat | Family | WcBJ | GC | CPM | GT |
|---|---|---|---|---|---|---|
| *Notorynchus cepedianus* | Coastal | Carcharhinidae | X | X | | |
| *Odontaspis ferox* | Coastal | Odontaspididae | X | X | | |
| *Parmaturus xaniurus* | Deep sea | Pentanchidae | X | | | |
| *Prionace glauca* | Pelagic/ Oceanic | Carcharhinidae | X | X | X | X |
| *Pseudocarcharias kamoharai* | Pelagic/ Oceanic | Pseudocarchariidae | X | | | |
| *Rhincodon typus* | Pelagic/ Oceanic | Rhincodontidae | | X | | |
| *Rhizoprionodon longurio* | Coastal | Carcharhinidae | X | X | X | X |
| *Somniosus pacificus* | Deep sea | Sominiosidae | X | | | |
| *Sphyrna corona* | Coastal | Sphyrnidae | | X | X | X |
| *Sphyrna lewini* | Coastal | Sphyrnidae | X | X | X | X |
| *Sphyrna media* | Coastal | Sphyrnidae | | X | X | X |
| *Sphyrna mokarran* | Coastal | Sphyrnidae | | X | X | X |
| *Sphyrna tiburo* | Coastal | Sphyrnidae | | X | X | X |
| *Sphyrna zygaena* | Coastal | Sphyrnidae | X | X | X | X |
| *Squalus suckleyi* | Coastal | Squalidae | X | | | |
| *Squatina californica* | Coastal | Squatinidae | X | X | | |
| *Triaenodon obesus* | Coastal | Carcharhinidae | X | X | | X |
| *Triakis semifasciata* | Coastal | Triakidae | X | X | | |

fisheries. Neritic species are concentrated especially in river mouths, estuaries, bays, and marine lagoons. Primary production in those environments provides the necessary nutrients and food year-round for the development of trophic chains composed of chondrichthyan species. The diversity and abundance of coastal shark species exhibit 'downs' and 'ups' between years because of various factors. Extended winter or summer seasons reduce primary productivity at sea, while decreasing rainfalls depress coastal nutrient levels, both of which affect communities and assemblages. Oceanographic phenomena like the El Niño and La Niña also impact coastal shark

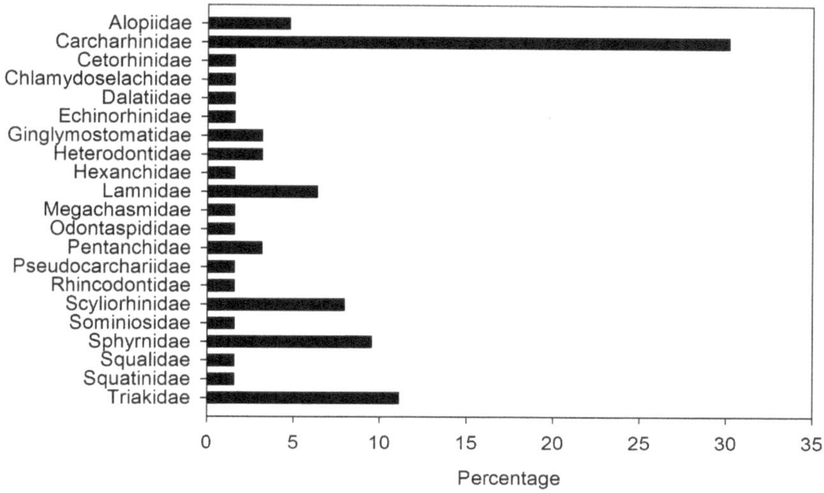


**Fig. 3** Percentages of species of shark in the Northeast Pacific Ocean by families.

populations, forcing species to occupy different areas or refuges. 'The Blob', an atypical warm mass of water that originated in the northeast Pacific off Alaska, affected the entire California Current system during 2015 and 2016. The thermal anomalies were higher than the 1998–99 El Niño and impacted the shark fauna along the entire west coast of Baja. Yields and catches of blue shark, *Prionace glauca*, dropped below the average during those years as the logbooks and observer program reported (INAPESCA unpublished data). There were impacts on the fauna in Sebastián Vizcaino Bay (SVB), which is a large bay characterized by high productivity located in the middle of the west coast of the Baja peninsula. It is home to a high diversity of elasmobranchs year-round (Cartamil et al., 2011). Juvenile blue sharks dominate by number and weight in the commercial catches year by year. During 2015 and 2016, blue shark landings diminished to a minimum level, whereas shortfin mako catches increased threefold. Blue sharks almost disappeared from the bay and found refuge in deeper waters. The occupation of SVB by the shortfin mako can be explained by its preference for the comparatively high sea temperatures created by 'The Blob' (INAPESCA unpublished data).

The second group of sharks present in the Pacific Ocean surrounding Mexico consists of oceanic and pelagic species, which are distributed in the northern and central Pacific regions. Those species are very sensitive to changes in oceanographic conditions, especially to SST (Sea Surface

Temperature). The three most important species are the blue shark, shortfin mako shark, and silky shark. They are highly migratory.

It is widely recognized that sharks play a key role in the transfer of energy between upper trophic levels within marine ecosystems (Bizzarro et al., 2017; Wetherbee and Cortés, 2004). There are apex predators such as the white shark *Carcharodon carcharias*. Also present are filter feeders such as the whale shark *Rhincodon typus*. Finally, there are suction-crushing species such as the horn shark *Heterodontus francisci* and opportunistic scavengers such as the peppered catshark *Galeus piperatus*. While a few large species >2 m in total length function as top predators, a greater number of species are mesopredators <1.5 m in total length that are prey for larger species (Ferretti et al., 2010). This is reflected by the high connectivity of sharks in food web models (Bascompte et al., 2005) and the limited effect of any single apex predator on a particular prey species (Ellis and Musick, 2007). In this sense, the impact of removing sharks from a marine ecosystem is complex and varied. On the one hand, the removal of apex predators can cause changes in prey behaviour and survival rates that result in trophic cascades or collapses of trophic levels below the apex predator, disrupting ecosystem functioning and biodiversity (Ferretti et al., 2010).

Apex predators can be key to preventing outbreaks of mesopredators and the consequent ecological, financial and social problems (Prugh et al., 2009). An example of this effect is given by Myers et al. (2007), who observed that overfishing of shark in the Atlantic Ocean led a population explosion of the Cownose Ray (*Rhinoptera bonasus*), which in turn reduced Bay Scallop (*Argopecten irradians*) populations to such low levels that a century-old scallop fishery was forced to close. Comprehending the ecological role and the position of each shark species within a food web is the first step to clarifying the dynamics of marine communities and the impact that each species has on trophic networks (Lucifora et al., 2012), which is critical to effective resource conservation and responsible exploitation.

Compared to teleost fishes, sharks are characterized by relatively conservative life-history traits (e.g. growth rate, fecundity) and are, thus, considered K-selected species (Cailliet et al., 2005; Stevens et al., 2000). Life-history traits of sharks vary widely, particularly depending on their reproductive traits: gestation period (from 2 to 42 moths), reproductive mode (egg-laying, live-bearing), egg hatching period (from 1 to 27 months), maternal investment (yolk-only, uterine milk, oophagosity, uterine cannibalism, placentation), fecundity (from 1 to 400 offspring), offspring size (from 20 to 1800 cm long), age at maturity (1.5 to >30 years), and longevity

(5 to >50 years) (Compagno, 1990; Cortés, 2000; Dulvy and Reynolds, 2002). Additionally, most sharks display slow growth, long gestation periods, late maturity, low fecundity and productivity and long-life cycles (Cailliet et al., 2005; Camhi et al., 1998; Cortés, 2000). These biological traits generally result in low reproductive potential and low capacity to increase population size after perturbation by a stressor. This has serious implications for shark populations, limiting their capacity to sustain fisheries and to recover from declines (Cailliet et al., 2005; Cavanagh and Gibson, 2007).

# 3. Trophic ecology of sharks off the Pacific Coast of Mexico by area

## 3.1 West coast of Baja California (WcBJ)

### *3.1.1 White shark (Carcharodon carcharias)*

The White Shark inhabits temperate waters of the Pacific, Atlantic, and Indian Oceans. Here, we build upon existing knowledge about the trophic ecology of the White Shark in the Mexican Pacific. Adult white sharks have been reported in Guadalupe Island and the central part of the Gulf of California (Galvan-Magaña et al., 2010; Hoyos-Padilla et al., 2016), while the juveniles have been frequently observed on the west coast of the Baja California Peninsula (Benson et al., 2018; Castillo-Geniz et al., 2016; Oñate-González et al., 2017; Santana-Morales et al., 2012).

### *3.1.2 White Shark Guadalupe Island*

Guadalupe Island is considered a 'hot spot' for White Sharks in the eastern Pacific. White sharks are present at the island between July and January, with August to December as the peak months (Domeier and Nasby-Lucas, 2007). Using ultrasonic telemetry to track white sharks, Hoyos-Padilla et al. (2016) were able to describe a differential use of the island by the different life-history stages. The data collected suggest that juveniles arrive at Guadalupe Island from nursery grounds on the mainland after they have reached at least 180 cm TL; then they remain around the island for several months (up to 14), potentially taking advantage of the diversity of prey. They remain close to the island throughout the day between the surface and 50 m depth in warm waters from 14 to 20 °C to feed on prey abundant in shallow waters, such as Bat Rays *Myliobatis californica* (Domeier and Nasby-Lucas, 2007). In addition, juveniles show patterns such as shallow dives during the night that could be related to feeding. There are several diurnal migrators present in

Guadalupe Island surface waters at night that are potential prey for juvenile White Sharks including squid (*Onychoteuthis banksi*, *O. borealjaponica*, and *Dosidicus gigas*), two types of mackerel (*Scomber japonicus* and *Auxis thazard*), sardines (*Sardinops caerulea*), flying fish (*Cypselurus californicus*), and anchovies (*Engraulis* spp.) (Gallo-Reynoso, 1994).

It has been argued that the distribution of large White Sharks at Guadalupe Island is potentially associated with foraging and the seasonal cycles of pinnipeds (Domeier, 2012). White sharks at Isla Guadalupe behead pups of Guadalupe fur seals and ambush Northern elephant seals in deep areas (Gallo-Reynoso et al., 2008; Hoyos-Padilla, 2009). Although seal carcasses have been observed floating at the surface, the predation event has not been witnessed. In addition, satellite (Domeier, 2012) and acoustic tracking (Hoyos-Padilla et al., 2016) data indicate that White Sharks routinely make daily dives to depths in excess of 100 m when around Guadalupe, though the reason for this remains a mystery.

Skomal et al. (2015) tested an autonomous underwater vehicle (AUV) for directly observing the behaviour, habitat use, and feeding ecology of White Sharks near Guadalupe Island. The four white sharks tracked during this study spent, on average, 80% of the time at depths >100 m and only 5% of their time at depths <25 m. Collectively, these observations suggest that predation events occur below the surface. In other areas, it has been established that White Sharks avoid the surface and remain at depths down to 50 m while near pinniped rookeries in autumn and winter. This is consistent with a silhouette-based ambush hunting strategy (Klimley, 1994). Skomal et al. (2015) observed White Sharks approaching, bumping, and biting the AUV at depths ranging from 53 to 90 m, thereby providing direct evidence of predatory behaviour at depth. These data suggest that White Sharks take advantage of great underwater visibility to search for seals in deep water adjacent to seal colonies where they ambush and disable pinnipeds and, perhaps, follow the carcass to the surface (Hoyos-Padilla, 2009). It is also possible that White Sharks take advantage of seals in Guadalupe Island before they go to their pupping grounds to give birth in California and Baja California or to their offshore migration to the west.

### 3.1.3 White Shark Gulf of California

According to Jaime-Rivera et al. (2014) the White Shark has an ecological role that links food webs in Pacific offshore and nearshore areas from Mexico. They found evidence that some sharks from Isla Guadalupe move into the Gulf of California, according to the isotopic contribution of the jumbo squid to dermal tissue. Large cephalopods are an important diet component of white sharks and are indicators of habitat use (Smale and Cliff, 2012); therefore, this species can be considered a main prey item there. Moreover, jumbo squid recruitment in the Gulf of California occurs during April and May and population size in April is estimated at ~136 million individuals (Morales-Bojórquez et al., 2012). This is the same season when most white sharks have been caught in the gulf (Galvan-Magaña et al., 2010).

### *3.1.4 White Shark Baja California Peninsula*

Although modern technology has elucidated interesting facts about adolescent and adult movements, almost nothing is known about juveniles. Until very recently the understanding of young-of-the-year biology was based largely on the incidental take of juveniles and pregnant females, and stomach content analyses from a relatively small number of individuals (Klimley, 1985a,b). Given the considerable shift in diet and differences in geographic location, inferences about juveniles based on adult behaviour are questionable (Dewar et al., 2004).

In Mexico, Santana-Morales et al. (2012) documented between 1999 and 2010 the incidental catch of 111 juvenile White Sharks during surveys of the artisanal and commercial fisheries along the Pacific coast of Baja California, and opportunistically collected stomach samples. Sharks ranged in size from 123 to 274 cm total length (TL). Most (79.8%) were young-of-the-year (YOY) ($\leq$175 cm TL) and the remainder were juveniles (>175–300 cm TL). Of the four documented fishing-gear types, bottom gillnets were used to capture 74.7% of the individuals, followed by drift gillnet (18.0%), artisanal seine net (4.5%), and surface longline (2.7%). Catch data indicate that the continental shelf is an important habitat for YOY and juveniles in the region, with a possible core nursery area in Bahia Sebastian Vizcaino, where 66 YOY specimens were captured. Stomach content analysis for 14 juveniles are consistent with previous studies showing that the species feeds primarily on bony fish and demersal elasmobranchs including *Thunnus* spp., *Scomber japonicus*, *Atractosion nobilis*, *Myliobatis californica*, *Mustelus* spp., unidentified Scombridae family., cephalopods (order Teuthoidea), crustaceans, and egg capsules of *Raja* spp. The latter prey in their stomach indicates that they feed in the coastal waters (Santana-Morales et al., 2012). These stomach contents observed by Santana-Morales et al. (2012) closely resemble the prey composition previously reported for juveniles by Klimley (1985a,b) in the west coast of North America. A 149.5-cm-TL White Shark was found with 12 of its own teeth in its stomach. The size range of these teeth (between 3 and 6.5-mm crown height) corresponds to its embryonic stage (Kabasakal and Özgür-Gedikoğ, 2008), suggesting that the shark was recently born during the summer months in coastal Baja California waters. Furthermore, Malpica-Cruz et al. (2013) attributed high $\delta^{13}C$ values of YOY white sharks to consumption of benthic fishes inside Bahia Sebastian Vizcaino. Tamburin et al. (2019) found that the isotopic composition of YOY white sharks supports the hypothesis that these sharks reside and forage in Bahia Sebastian Vizcaino and/or similar inshore regions along Baja California for extended timeframes. Yet they may move between the inshore regions in both northern and southern Baja.

## 3.2 Pelagic sharks from the western coast of Baja California

### *3.2.1 Blue Shark (Prionace glauca)*

The blue shark *Prionace glauca* is the main species caught in the western coast of Baja California. It is the most abundant and widespread shark (Ramirez-Amaro et al., 2013). Despite its importance, there are very few studies on its feeding habits. Hernández-Aguilar et al. (2016) reviewed 368 samples from the artisanal fishery in the western coast of Baja California Sur. The trophic spectrum of *P. glauca* was 13 cephalopods, 7 fish, 3 crustaceans, 1 macroalgae, 1 bird, and 1 elasmobranch. Using the Index of Relative Importance (IRI) the most important prey were Red crab (*Pleuroncodes planipes*), followed by three species of squid: *Gonatus californiensis*, *Ancistrocheirus lesueurii*, and *Haliphron atlanticus*. They conclude that blue shark is a predator that feeds mostly on squid, showing considerable vertical migrations given that its prey have epipelagic, mesopelagic, and benthic habitats. Markaida and Sosa-Nishisaki (2010) analysed 893 blue shark stomach contents from the western coast of Baja California, which indicate that this shark consumes mainly cephalopods of the following species: *Histioteuthis heteropsis*, *Gonatus californiensis*, *Argonauta* spp., and *Vampyroteuthis infernalis*. The pelagic crustacean red crab (*Pleuroncodes planipes*) was one of its main prey. Using stable isotopes analysis indicates that blue shark can feed in coastal and oceanic waters (Maya-Meneses et al., 2016; Polo, 2013; Polo-Silva et al., 2012). Also Méndez Da Silveira (2015) found in the blood and muscle of blue sharks isotopes from the tip of Baja California values that indicate the trophic habitat of this shark is dominated by oceanic waters. The trophic level of blue shark in Baja California Sur waters is of 4.05, which indicate as tertiary consumer (Hernández-Aguilar et al., 2016).

### 3.2.2 Silky Shark (Carcharhinus falciformis)

A total of 263 stomach contents were analysed. The index of relative importance (IRI) showed that silky sharks fed mainly on red crabs *Pleuroncodes planipes* and jumbo squid *Dosidicus gigas* followed by chub mackerel *Scomber japonicus* (Cabrera Chavez-Costa et al., 2010). Also, differences exist between the main prey of juveniles and adults. The smaller silky sharks fed mainly on *D. gigas*; whereas the larger specimens consumed more *S. japonicus*. Cabrera Chávez-Costa (2003) used stable isotopes in silky shark muscle to distinguish between the trophic habits of males and females, and between juveniles and adults. The results indicate that individuals of both sexes feed in oceanic waters. Also, Méndez Da Silveira (2015) conducted isotopic analyses of silky sharks in the southern region off Baja California which indicates that the silky shark feed in coastal and oceanic habitats.

### 3.2.3 Smooth Hammerhead Shark (Sphyrna zygaena)

Little information exists worldwide on trophic habits of this species. In two areas of the western coast of Baja California, studies indicate this species feeds mainly on fish species in one area: *Mugil cephalus* (30%), *Sardinops caeruleus* (23%), *Brotula* spp. (19%), *Synodus evermanni* (1.4%); and cephalopod species in another area, including *Pholidoteuthis boschmai* (28%), *Onychoteuthis banksii* (36.2%), *Ancistrocheirus lesueurii* (27.1%), *D. gigas* (28.7%), and *Stenoteuthis oualaniensis* (3.7%) (Ochoa Diaz, 2009).

Also, stable isotopes in muscle were analysed for this species off the western coast of Southern Baja California. This indicates that females and males predate within different trophic habitats. Males tend to be more oceanic, compared to females whose carbon source indicates they are more coastal (Ochoa Diaz, 2009). Méndez Da Silveira (2015) examined stable isotope values from animals around the tip of Baja California, where this species feeds in oceanic waters.

## 3.3 Coastal sharks of the western coast of Baja California

### 3.3.1 Tope Shark (Galeorhinus galeus)

The Tope Shark was classified as a high trophic level piscivorous predator in this area, feeding mainly on the teleost fish *Synodus* spp. (51.68%), *Sardinops sagax* (11.66%), and *Scomber japonicus* (11.2%), and in smaller proportions on *Octopus bimaculatus* (6.67%) (Di Filippo, 2018). Stable isotope analyses of this species found values of $\delta^{13}C$ and $\delta^{15}N$ which indicated that juveniles and females are associated more with coastal environments, compared to adults and males which feed more in oceanic waters. However, there is a trophic difference in the usage of resources within the stages of development, but not between sexes. An increase in trophic level was also observed as organisms increased in size, which may be related to an increase in energy demand by larger organisms, or a strategy to avoid competition and/or predation (Di Filippo, 2018).

### *3.3.2 Sicklefin Smooth-hound (Mustelus lunulatus)*

This dogfish is an important species in the artisanal fisheries of the western coast of Baja California Sur. Their diet includes 3 species of cephalopods, 29 crustacean species, and 9 fish species. According to the relative prey index of relative prey (PSIRI), the red crab *Pleuroncodes planipes* was the most important prey species (39.94%) followed by the crustacean *Hemisquillia californiensis* (19.46%) in the Northwest Baja California Sur; while in the tip of Baja California the crab *Platymera gaudichaudii* was the most important (25.87%), in addition to two species belonging to the genera *Achelous* spp. (4.43%) and *Portunus* spp. (4.49%). This indicates that they reside in trophic level that ranged between 3.7 and 3.85 in both areas (Martínez, 2018). The isotopic analysis from muscle samples of *M. lunulatus* in the northwest of Baja California Sur indicate that this species feed mostly on crustaceans in benthic areas from deep sea. Within each category (sex, stage of maturity, season) no significant differences were found in the isotopic values.

### *3.3.3 Brown Smooth-hound (Mustelus henlei)*

An analysis of 185 stomachs indicated that *Mustelus henlei* feed on 43 prey species, including cephalopods, crustaceans, and fishes. Using the Index of relative importance (IIR), the cephalopod *Lolliguncula diomedeae* (57.31%) and red crab *Pleuroncodes planipes* (32.29%) were the most important prey. The females feed mainly on *L. diomedeae* (53.74%), *P. planipes* (34.71%), and *Plathymera gaudichaudii* (2.01); whereas males feed on *L. diomedeae* (62.08%) and *P. planipes* (28.01%). This shark is a specialist species with a very narrow range of dietary diversity (Espinoza, 2012). Méndez Da Silveira (2015) analysed stable isotopes in muscle of this shark species which indicates that they inhabit oceanic waters.

### *3.3.4 Swell Shark (Cephaloscyllium ventriosum)*

The food habits of *Cephaloscyllium ventriosum* are known through stomach contents analysis of 86 organisms from the northwest Baja California Sur. The percentage of prey-specific index of relative importance (%PSIRI) indicates that two most important prey were the crustacean *Hemisquilla ensigera californiensis* (31.32%) and the cephalopod *Octopus bimaculatus* (22.31%) (Baro, 2016).

### 3.3.5 Mako Shark (*Isurus oxyrinchus*)

The catch of this species by the artisanal fishery of the west coast of Baja California Sur is confined mainly to neonates and juveniles. This coastal area is a nursery area for this species. A total of 269 stomachs were examined by Velasco (2005), from which 31 prey were identified. The most important fishes in the diet were *Prionotus albirostris* (65% IRI) and *Scomber japonicus* (5%), while for cephalopods (27%), the squids *Dosidicus gigas* (23%) and *Ancistrocheirus lesueurii* (4%) were the most important prey. The juvenile mako shark was classified as specialist predator. By sex, the males had a diet less selective than females. The prey consumed by adults were mainly pelagic and of bigger size, whereas juveniles predate mainly on benthic prey such as *P. albirostris* (Velasco, 2005).

Using isotopic values of nitrogen ($\delta^{15}N$), three trophic levels were found to be evident in different life stages. A higher trophic level was in juveniles ($\mu = 18.10‰ \pm 0.82$) and adults ($\mu = 18.72‰ \pm 0.57$). The most important prey was *S. japonicus* ($\delta^{15}N$: 17.36‰) and *P. albirostris* ($\delta^{15}N$: 16.73‰), with trophic position similar to the juvenile Mako Shark. Also the prey *Dosidicus gigas* ($\delta^{15}N$: 15.68‰) had a similar trophic level compared to Mako Shark juveniles, but a lower trophic level in relation to adults. The prey *Mugil cephalus* ($\delta^{15}N$: 11.02‰) and *P. planipes* ($\delta^{15}N$: 12.30‰) had low values. Using carbon isotopes ($\delta^{13}C$), the most important prey in oceanic waters was *Coryphaena hippurus* (−16.11‰), *Scomber japonicus* (−16.92‰), *P. albirostris* (−17.53‰), and *D. gigas* (−17.94‰). The less important prey were *Mugil cephalus* ($\delta^{13}C$: −10.01‰) from coastal waters and *Pleuroncodes planipes* ($\delta^{13}C$: −19.03‰) from oceanic waters. The combination of results using two methods (stomach contents and Carbon and Nitrogen isotopic values) discriminated the differences between juveniles and adults. The western coast of Baja California Sur is an important feeding area for Mako Shark juveniles (Velasco, 2005).

### 3.3.6 Horn Shark (*Heterodontus francisci*)

The dietary habits of the horn shark were examined in the Northwestern area of Baja California Sur during spring, summer, and fall of 2014 (Cortés, 2015). A total of 78 stomachs were collected. According to the Index of Relative Importance (%IRI), the most important prey were anomurans (66%), cephalopods (7.2%), lobsters (4.7%), fish (4.2%), and sea urchins (2.3%). The main prey were the anomuran *Blepharipoda occidentalis* (65.2%), the octopus *Octopus bimaculatus* (5.4%), the lobster *Panulirus interruptus* (4.7%), and the sea urchin *Echinometra vanbrunti* (2.6%). Females fed primarily on *B. occidentalis* (69.6%), *O. bimaculatus* (9.5%), and squid remains (3.6%); while males feed on *B. occidentalis* (33.3%), *P. interruptus* (8.7%), and the fish *Calamus brachysomus* (5.3%). Juveniles fed primarily on *B. occidentalis* (22.9%), the squid *Doryteuthis opalescens* (6.5%), and the anomuran *Lophomastix diomedeae* (2.5%); while adults fed on *B. occidentalis* (66.8%), *O. bimaculatus* (6.9%), and *P. interruptus* (6.2%). According to trophic niche breadth *H. francisci* is a specialist predator. This species was classified as a tertiary consumer with trophic position of 4.06 (Cortés, 2015).

Segura et al. (1997) found, in 108 stomachs from San Ignacio Lagoon, the following main prey using IRI: the gastropods (*Lucapinella milleri* and *Calyptraea* spp.), crustaceans (*Callinectes bellicosus*, *Penaeus* spp. and isopods), pelecypods, sipunculids (*Sipuncula nodosus*), cephalopoda (*Octopus* spp.), and the fish (*Syngnathus auliscus*). This horn shark was a third order consumer that consumed a variety of benthic species.

Rodríguez (2016), using the stable isotopes of $\delta^{13}C$ and $\delta^{15}N$ in 71 muscles from the Northwest Baja California Sur, indicates that *H. francisci* feed mostly in oceanic habits. There were no significant differences between males and females, or between adults and juveniles. Using the SIBER method the females (1.99) have a higher trophic niche breadth than males (1.62), and both adults (1.22) and juveniles (2.31) are generalists. The trophic position being $3.31 \pm 0.29$ and indicating that *H. francisci* is a third order predator.

### *3.3.7 Pacific Angel Shark (Squatina californica)*

There is little information on feeding habits of this species in the western coast of Baja California Sur other than the stable isotope analysis completed by Valdivia (2013), which indicates that *S. californica* feeds in coastal areas and predates within the benthic habitat. No significant differences were found in the isotopic values between males and females, nor between juveniles and adults. This indicates that they share the same foraging areas.

## 3.4 Gulf of California (GC)

### *3.4.1 Whale Shark (Rhincodon typus)*

The Gulf of California (GC) is considered a transitional zone where different currents converge (Kessler, 2006), which supports high productivity, rich biodiversity, and high levels of endemism (Brusca, 2010). The GC encompasses different marine and coastal ecosystems that support the feeding habits of many benthic and pelagic shark species. In the pelagic area, the Whale Shark (*Rhincodon typus*), which is a cosmopolitan species that lives in both epipelagic and neritic waters in tropical and warm-temperate regions around the world (Compagno, 2001), is common. In the eastern Pacific it is recorded from northern California to the northern coast of Chile (Compagno, 2001; Ebert et al., 2004), including the Pacific coast of Mexico and the Gulf of California (Wolfson, 1987). In the Gulf of California, Whale Shark aggregations are recorded near Cabo San Lucas (Wolfson, 1987), Bay of La Paz (Clark and Nelson, 1997; Ketchum et al., 2013; Ramírez-Macías et al., 2012) and Bahía de Los Angeles (Nelson and Eckert, 2007; Ramírez-Macías et al., 2012), as well as around offshore seamounts, particularly at El Bajo Espíritu Santo and Gorda Banks (Ketchum, 2003; Ketchum et al., 2013; Klimley and Butler, 1988; Ramírez-Macías et al., 2017). One of the first studies to understand the movements and migratory patterns of Whale Sharks was carried out in the Gulf of California (Eckert and Stewart, 2001). This study, and more recent ones, found that juvenile Whale Sharks move north and south within coastal areas of the Gulf of California, whereas adults move more in oceanic waters (Eckert and Stewart, 2001; Ramírez-Macías et al., 2017).

The Whale Shark is an elusive fish, and little was known about its ecology in the Gulf of California until the last 20 years. Whale Sharks show seasonal, interannual, and spatial segregation by sex and size in the southern Gulf of California (Eckert and Stewart, 2001; Ketchum, 2003; Ketchum and Lozano, 2000; Ketchum et al., 2013). Large aggregations of juveniles have been observed inshore within the southern portion of the bay, mainly juvenile males, whereas, offshore aggregations in deeper waters are composed of adults, mostly females (Ketchum, 2003; Ketchum et al., 2013). Whale shark prey abundance and preference have been described in coastal regions (Clark and Nelson, 1997; Hacohen-Domené et al., 2006; Ketchum et al., 2013; Nelson and Eckert, 2007, Taylor, 2007), as well as associations of the species with hydrographic features of lagoon and shelf waters (Wilson et al., 2001). Whale sharks associate with particular sea surface temperature in La Paz Bay, Mexico (19.6–29.8 °C) (Ketchum et al., 2013). However, information about habitat preferences of Whale Sharks and the characterization of their habitat is far from complete and studies on distribution associated with concurrent hydrographic and biological features are still lacking in many regions. In the Gulf of California, aggregations of Whale Sharks occur at bays and offshore seamounts (Clark and Nelson, 1997; Ketchum et al., 2013; Klimley and Butler, 1988; Nelson and Eckert, 2007), which may be important feeding and nursery grounds for the species.

Whale Shark habitat in the Bay of La Paz can be characterized by the dynamic physical and biological characteristics of the bay's waters (Ketchum et al., 2013). Hence, according to the variables analysed, whale shark habitat in the bay may be subdivided into: (1) coastal areas dominated by inshore processes such as tidal mixing and nutrient input from a coastal lagoon; (2) mixed areas with both inshore and offshore dynamics; and (3) oceanic areas dominated by offshore phenomena like the presence of thermal fronts and a sharp thermocline (Ketchum, 2003). The coastal south and southeastern areas of the Bay of La Paz harbours the highest abundance of whale sharks, especially juveniles with a density of ~4 sharks/km$^2$ (Ketchum et al., 2013). Such density of sharks is very similar to the one recorded for Ningaloo Reef in Australia (4 sharks/km$^2$; Taylor, 1994), but much smaller than the 4 sharks/ha recorded for Gladden Spit in Belize (Heyman et al., 2001). The southern bay is influenced by its location near a current convergence zone off the narrowest portion of the Mogote sandbar (Jiménez-Illescas et al., 1997) and lies within the strong tidal currents off Punta Prieta (Ketchum et al., 2013). These two features promote intense secondary production during most of the year, which supports a high biomass and a zooplankton community structure composed mainly by copepods in this area (Ketchum et al., 2013). The latter had been also observed by previous studies (Palomares-García, 1996). In the southeastern bay, the tidal flows are dominant (Obeso-Nieblas et al., 2002), and act in conjunction with the outflow of nutrients from a coastal lagoon (Ensenada de La Paz) and the local topography (Signoret and Santoyo, 1980) to have a strong influence on the density and composition of zooplankton (Ketchum et al., 2013). Hence these two coastal areas provide highly favourable environmental conditions and preferred prey for juvenile Whale Sharks. In the northwestern part of the bay, a mixed area, the abundance of juvenile sharks was lower and adult sharks occurred more frequently here than in the south (Ketchum et al., 2013).

The Whale Shark is a filter feeder that feeds on a variety of prey, from zooplankton to small fish and squid (Last and Stevens, 1994). In contrast to other filter-feeder elasmobranchs such as the Basking Shark, *Cetorhinus maximus*, and the Manta Ray, *Mobula birostris*, Whale Sharks can capture larger prey by active suction filtering behaviour; however, individuals depend on dense aggregations of planktonic and nektonic prey (Colman, 1997), such as copepods, euphausids, sardines, and anchovies (Compagno, 1984). Juvenile Whale Sharks in the Gulf of California feed by active and stationary (or vertical) suction, and by ram-filtering (or passive feeding) between active and stationary bouts (Ketchum et al., 2013; Nelson and Eckert, 2007). For juvenile sharks, active suction is associated with dense copepod patches (70–85%), stationary suction with more sparse copepods (20–40%), and ram-filtering with the lowest concentration of copepods (<25%) in Bay of La Paz and Bahía de Los Angeles (Ketchum et al., 2013; Nelson and Eckert, 2007). The most abundant copepods found in the Bay of La Paz were *Acartia sp.*, *Undimula sp.* and *Corycaeus sp.* (Hacohen-Domené et al., 2006).

### 3.4.2 White Shark (Carcharodon carcharias)

The White Shark is also found in the Gulf of California. Galván-Magaña et al. (2010) found 38 records of White Shark captures and reliable sightings for the Gulf of California. The main locations of occurrence were El Golfo de Santa Clara (13 records), followed by Santa Rosalia (4), and San Pedro Martir and San Pedro Nolasco Islands (3 each). Presence of juveniles (≤300 cm total length) was highest from January to May (10 records) and mostly restricted to the upper Gulf, mainly in the relatively shallow waters off the fishing town of El Golfo de Santa Clara, and this is probably associated with foraging. Juveniles apparently show a dietary preference for benthic and coastal fishes (Klimley and Nelson, 1984; Tricas and McCosker, 1984). The northern Gulf of California is habitat for seasonally large schools of medium- to large-sized fish such as the endemic corvina (*Cynoscion othonopterus*) and totoaba (*Totoaba macdonaldi*), which could be consumed by juveniles. Records of subadults and adults (300–400 cm TL) were more common from December to May (eight records), but less frequent from June to October (five records). The stomach contents were identified for some adults and consisted of bottlenose dolphins (*Tursiops truncatus*), unidentified whale remains, and California sea lions (*Zalophus californianus*).

### 3.4.3 Smooth Hammerhead Shark (Sphyrna zygaena)

Another pelagic shark species known to inhabit the GC is the Smooth Hammerhead Shark. This is a coastal-pelagic and semi-oceanic species that occurs on the continental shelf, to 200 m depth (Ebert, 2003). One possible pupping ground and nursery area for this species is the northern GC. In general, the trophic spectrum of *S. zygaena* in the GC it is represented by 23 prey species, where according with the index of relative importance (IRI) mesopelagic cephalopods like *Dosidicus gigas* (75%), *O. banksii* (11%),

*S. oualaniensis* (8%), and *A. lesueurii* (4%) are the main prey species in its diet (Ochoa Diaz, 2009). The Smooth Hammerhead shows ontogenetic differences based on stable isotopes (Ochoa Diaz, 2009), where females seem to feed more on the coast and the males feed more in oceanic areas and mainly on epipelagic fish. Adults of both sexes consume mesopelagic squids and juveniles consume predominantly benthic prey species.

### *3.4.4 Scalloped Hammerhead Shark (Sphyrna lewini)*

A related species that demonstrates similar feeding behaviour to the Smooth Hammerhead Shark is the coastal and semi-oceanic pelagic Scalloped Hammerhead Shark. Adults occupy the southern Gulf of California from late spring to late fall and may arrive coincident with an assemblage of pelagic fishes coming from the tropical eastern Pacific Ocean (Klimley and Butler, 1988). This species spends daytime hours around offshore seamounts where they form schools before making nighttime migrations away to forage in the pelagic environment (Klimley and Nelson, 1984). Female hammerheads dominate the large schools that exist at seamounts and offshore islands (Klimley and Nelson, 1981). They compete among each other for a central position within the schools, where males pair with the largest dominant females (Klimley, 1985a,b). The male copulates with the female, as the two descend in the water column from a shallow depth, and this is likely to occur over deep water near the schooling sites (Salinas-de-León et al., 2017). The females migrate offshore at an earlier age than males, and by doing so increase their rate of growth leading to a larger body mass than males prior to reaching sexually maturity (Klimley, 1987). Supporting this hypothesis was the migration of an immature Scalloped Hammerhead equipped with an archival tag from coastal waters, apparent in the recorded shallow dive depths during both day and night, to an offshore seamount, evident from shallow daytime depths and deep nighttime depths (Hoyos-Padilla et al., 2014). The dive records on the archival tag are consistent with the females remaining in schools to shallow waters above seamounts, and their deep dives when moving away from the seamount into deep water to feed at night. Individuals have been observed to make nighttime migrations to a distance of 20 km from a seamount and return to it the following day (Klimley, 1993). The schools appear to depart from the seamount in response to upwelling of cool, deep waters, only to return with warmer waters (Klimley and Butler, 1988).

Juvenile Scalloped Hammerheads inhabit the shallow coastal bays of the eastern coast of Baja California (Klimley, 1987) and western coast of the

Mexican mainland (Torres-Huerta, 2004). Many species of sharks use coastal–estuarine environments as nursery grounds. These areas offer abundant food resources to the young, and better protection from predators than would be afforded to them in the pelagic environment (Clarke, 1971; Simpfendorfer and Milward, 1993). Bush and Holland (2002) mentioned that sharks use nursery areas where there are accessible food items (i.e. catchable prey), and high prey abundance (i.e. nutrient-rich habitat). Based on isotopic mixing models, Torres-Rojas et al. (2013) suggest that *S. lewini* feed on prey present in the Gulf of California, therefore this region could be considered a nursery area with abundant food for this shark species.

The offshore movement of female Scalloped Hammerhead at an earlier age than males and the use of coastal areas by juveniles in the southern Gulf of California results in different diets between sexes and over ontogeny. To illustrate this, prey species, grouped according to habitat, are shown in pie-shaped diagrams for both sexes (Fig. 4). In order to detect ontogenetic divergence, the diet is further separated for sharks of less than or equal to and >160 cm TL. This division was chosen because males did not enter the offshore schools until within this size class while females in the 100 cm size class were seen in the offshore schools (Fig. 5). As might be expected from their early offshore movement, females in the <160 cm size class fed on a higher percentage of pelagic prey than did males of similar sizes. Mesopelagic prey formed 27.5% and epipelagic prey 5.5% of the female total IRI, while such prey composed only 18.1% and 3.6% of the male total IRI. Furthermore, the diet of the females was composed of only 15.1% benthic prey in comparison to 40.9% of the IRI of male sharks.

A subsequent analysis of the feeding habits of juvenile Scalloped Hammerheads also identified differences suggesting that males and females occupy different habitats based on their stomach contents (Aguilar-Castro, 2003). In females a 44.6% IRI comprised three pelagic squid (*Abraliopsis affinis*, *Dosidicus gigas*, and *Onychoteuthis banksia*) and one pelagic fish (*Scomber japonicus*), whereas in males a 33.0% IRI was composed of only one pelagic squid (*Abraliopsis affinis*) and one pelagic fish (*Scomber japonicus*). The early offshore movement of females was found to result in an increase in their predatory success while similar size males, which remained inshore, did not increase their predatory success (Klimley, 1987). It is suggested that early offshore migration of females results in more rapid growth, allowing females to reach maturity at the larger size necessary to support embryonic young at a similar age to males, matching the female reproductive lifetime to that of the male (Hoyos-Padilla et al., 2014).

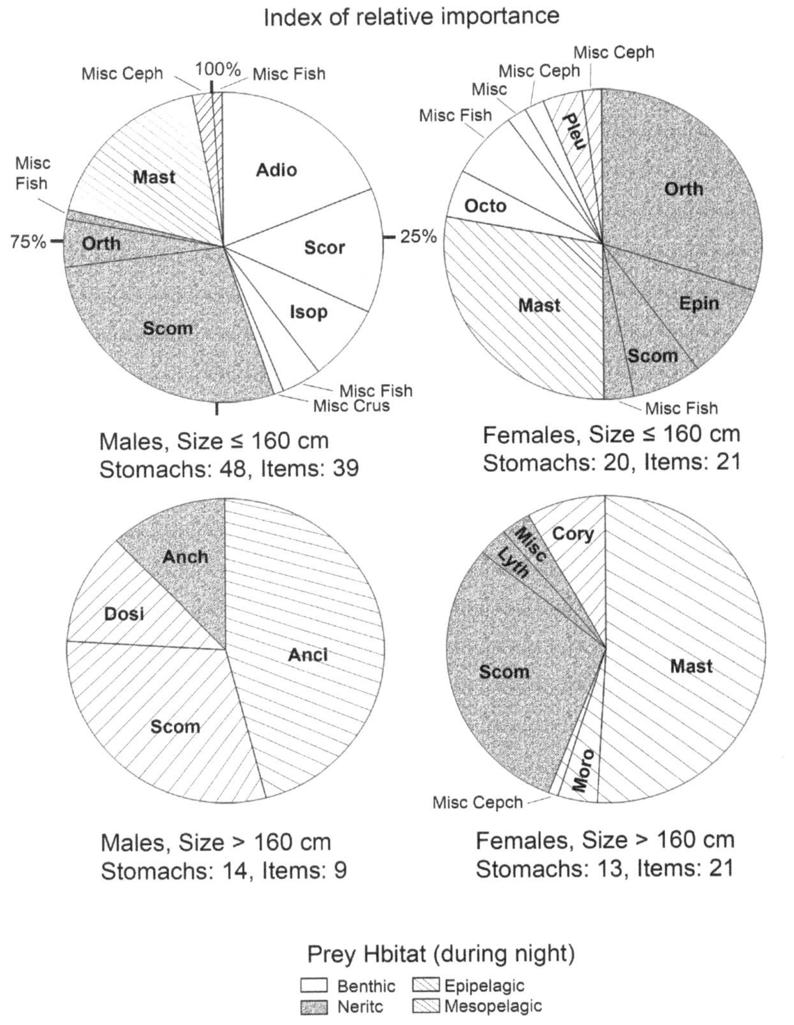


**Fig. 4** Relative importance of prey species for male and female Scalloped Hammerhead Sharks < and >160 cm captured by fishermen from May to October 1978–81 from Juncalito (25° 48′N, Il1° 18′W) to San Jose del Cabo (23° 03′ N, 109° 39′ W). Adio = *Adioryx suborbitalis*, Scor = *Scorpaena sonorae*, Isop = Isopoda, Scorn = *Scomber japonicus*, Orth = *Orthopristis inornatus*, Mast = *Mastigoteuthis* sp., Epin = *Epinephalus* sp., Octo = *Octopus* sp., Pleu = *Pleuroncodes planipes*, Anci = *Ancistrocheirus lesueuri*, Scomb = *Scomberomorus sierra*, Dosi = *Dosidicus gigas*, Anch = *Anchoa* sp., Moro = *Moroteuthis robustus*, Lyth = *Lythrulon flaviguttatum*, and Cory = *Coryphaena hippurus*. *Taken from Klimley, A.P. 1987. The determinants of sexual segregation in the scalloped hammerhead, Sphyrna lewini. Environ. Biol. Fishes, 18, 27–40.*

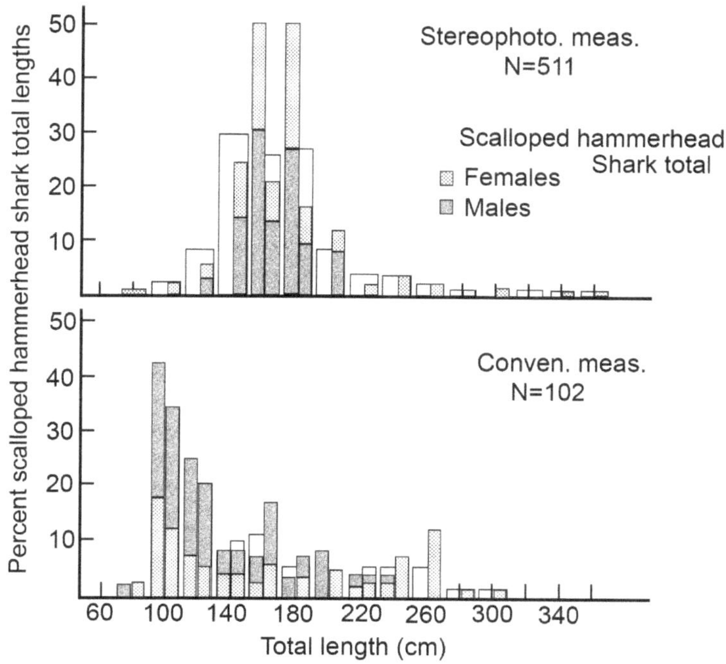


**Fig. 5** Percentages of lengths in different classes of Scalloped Hammerhead Sharks either measured by stereocamera in offshore schools (top) or by tape measure (bottom) from catches by fishermen in the Gulf of California. Cross-hatched bars are male, stippled bars female, and clear bars pooled frequencies. *Taken from Klimley, A.P. 1987. The determinants of sexual segregation in the scalloped hammerhead, Sphyrna lewini. Environ. Biol. Fishes, 18, 27–40.*

In the case of the feeding habits of juvenile hammerheads, the isotopic composition of individuals captured in the southeastern GC was similar to that of other predators that consume coastal benthic species (Blanco-Parra et al., 2012), where the main prey species consumed by smaller sharks (<100 cm) are squids. On the other hand, those individuals of >100 cm consumed more fish, which indicated changes in their feeding behaviour, but the lack of differences in the stable isotope of carbon suggests that juvenile populations of *S. lewini* did not differ in foraging habits (Torres-Rojas et al., 2013). Therefore, the presence of *S. lewini* in different ontogenetic stages in the GC is probably due to this shark using benthic coastal areas as feeding grounds.

### 3.4.5 Pacific Angel Shark (Squatina californica)

In the coastal areas of the GC, benthic shark species are common. The Pacific Angel Shark (*Squatina californica*), which is a benthic elasmobranch, is commonly found in soft bottom habitats, including shallow bays and estuaries (Ebert, 2003). In the GC, the Pacific Angel Shark matures at 75.6 cm for males and 77.7 cm for females (Romero-Caicedo et al., 2016). It has a seasonal presence, indicating a migration during the summer associated with reproductive activities (Galvan-Magaña et al., 1989). According to the IRI, the most important prey species of *S. californica* are the jack *Decapterus macrosoma* (47.5%), the daisy midshipman *Porichthys analis* (15.9%), the lizardfish *Synodus evermanni* (8.0%), and the crustacean *Sicyonia penicillata* (8.0%) (Escobar-Sanchez et al., 2006, 2011).

In the coastal areas of the GC small sharks are commonly captured, also called *cazón*, where the presence of the genus *Mustelus* and *Rhizoprionodon longurio* stand out. *Mustelus* comprises about 25 species of small benthic sharks that inhabit tropical and temperate waters on the continental shelves of all oceans. There are two principal species in the GC, *M. californicus* and *M. lunulatus*, which are distributed in temperate-warm and tropical seas. The former is distributed from northern California (US) to the coasts of Colima (Mexico) and the latter from northern California to southern Peru, including the Gulf of California (Compagno, 1984). One of the main differences between these shark species is the size at maturity. Females of *M. californicus* mature between 78 and 80 cm TL and males between 72 and 74 cm. In contrast, females of *M. lunulatus* mature between 94 and 99 cm and males between 89 and 94 cm. In the northern GC the feeding habits of these two species were analysed. The trophic spectrum was represented by 14 prey species for *M. californicus* and 11 prey species for *M. lunulatus*, the most important prey in both shark species was the crustacean *Squilla bigelowi* (Mendez-Loeza, 2004). This crustacean species is from benthic habitats, which distributed throughout the Gulf of California, being abundant in the northern part and, possibly, is the stomatopod species with the highest density of organisms in the area (Hendrickx and Salgado-Barragán, 1991). Ontogenetic differences in feeding behaviour were reported for both *Mustelus* species. According to Ellis et al. (1996), the differences in diet are controlled by factors such as shark size, structure of the mouth and dentition, and shark distribution in relation to its potential prey and other predatory species. Therefore, the dietary differences between shark species are probably associated with physiological differences.

### 3.4.6 *Pacific Sharpnose Shark (Rhizoprionodon longurio)*

The Pacific Sharpnose Shark (*Rhizoprionodon longurio*) is a small coastal species that lives over soft muddy and sandy bottoms of the continental shelf of the eastern tropical Pacific. *R. longurio* shows marked seasonal movement patterns in the GC (Marquez-Farias et al., 2004) and the Gulf seems to be an important pupping area (Bizzarro et al., 2000). The feeding habits of *R. longurio* were described by Alatorre-Ramirez et al. (2013) who mentions that this shark species has a high feeding activity overnight on benthic and coastal-pelagic prey species, implying that individuals makes vertical migrations in the water column.

Some studies have been reported that *R. longurio* segregates by sex and size in the GC (Castillo-Geniz, 1990), which could result in different prey being consumed by sharks of different sex and maturity stages. Castillo-Geniz (1990) found the Serranidae fish (*Diplectrum* spp.) and Family Muraenidae as main prey of this shark. This ontogenetic diet variation is associated with prey-predator relationships (Abrams, 2000). Based on the IRI, adult females consumed a higher amount of benthic species, while males consumed a higher amount of epipelagic fish. On the other hand, juvenile male sharks consumed mostly benthic prey, while juvenile females consumed mostly epipelagic prey, which confirms the ontogenetic segregation of this shark species (Alatorre-Ramirez et al., 2013; Castillo-Geniz, 1990; Castillo-Geniz et al., 1996).

## 3.5 Central Pacific Mexican (CPM)

Despite the richness of shark species in the Central Pacific Mexican (CPM), very limited information about the trophic ecology of sharks, and thus their position within the food web and role in the ecosystem, exists for the region. The only available studies in the CPM are those from Barajas-Calderón (2015, 2018) on the feeding habits and trophic ecology of the Silky Shark, *Carcharhinus falciformis*, from the coasts of the State of Jalisco. Some studies of the food habits of the Scalloped Hammerhead Shark, *Sphyrna lewini*, have also been undertaken in the state of Nayarit and Southern Sinaloa at the entrance of the Gulf of California (Hernández-Corona, 2011; Rentería-Bravo, 2016; Rentería-Bravo et al., 2019), a transitional oceanographic zone between the CPM and the GC (Lavín and Marinone, 2003).

Barajas-Calderón (2015) determined, based on analysis of the contents of 110 stomachs, that the main prey of juvenile *C. falciformis* in the CPM according to the index of relative importance were the cephalopod,

*Dosidiscus gigas* (%IRI = 32.1), followed by teleosts (%IRI = 24), the squid, *Ancistrocheirus lesueuri* (%IIR = 15.1) and the swimming crab *Portunus xantussi affinis* (%IRI = 11.3). Some results of this study indicated a sexual segregation of the population in this region, but an overlap in the diet of both sexes at some size intervals contradicted such conclusion. The species was found to tend to be a specialist predator, feeding on the most abundant species in the region, except at the youngest stages (94–121 cm of total length, TL) when it is a generalist.

The analysis of stable isotopes (nitrogen, $\delta^{15}N$ and carbon, $\delta^{13}C$) as well as the content of 168 stomachs of *C. falciformis* (60–229 cm of TL) caught by the artisanal fishery during 2012–16 in Jalisco showed no differences in the diet by sex or season of the year (Barajas-Calderón, 2018). A total of 33 prey taxa were identified of which, according to the prey-specific index of relative importance (%PSIRI), the Pacific thread herring, *Opisthonema* spp. (%PSIRI = 13.5%), *Portunus xantusii affinis* (%PSIRI = 12.7%) and *A. lesueurii* (%PSIRI = 8.8%) were the most important prey species in the diet. Stable isotope proportions determined for 63 muscle samples and 17 samples from the most important prey species demonstrated a lack of difference between the isotopic niches of both sexes and among sharks across seasons. However, significant differences were found in both isotopes by maturity stage, indicating different feeding areas and prey of different trophic levels. The $\delta^{15}N$ mean value was 16.37, while the mean value for $\delta^{13}C$ was 17.05 with the relationship C:N of 3.4. The isotopic mixing model indicated that *A. lesueurii* was the most important prey. The trophic level calculated with both methodologies turned out very similar, 4.4 with the stomach contents and 4.1 with stable isotopes analysis.

Analysis of the stomach contents of 177 *S. lewini* caught from 2007 to 2017 in Nayarit and Southern Sinaloa indicated that the diet of females and males was not significantly different, according to the ANOSIM test ($R = 0.004$, $P = 0.44$). Additionally, regardless of size group (<100 and >100 cm of TL) teleosts were the main prey. However, as in previous studies further north into the GC (Torres-Rojas et al., 2006), it was found that the importance of teleosts decreased with size (%IRI = 93 and 77, respectively). The second most important preys in sharks <100 cm of TL was crustaceans (%IRI = 5) followed by cephalopods (%IRI = 2), whereas in sharks >100 cm of TL cephalopods presented a larger importance (%IRI = 23) and crustaceans were absent in their diet (Rentería-Bravo, 2016; Rentería-Bravo et al., 2019). The change in the diet in relation to shark size has been related to movements towards oceanic areas during the night to feed, returning to the coast during the day (Smale and Cliff, 1998; Torres-Rojas et al., 2006).

The correlation between the stomach contents and the age of sharks from Nayarit and Southern Sinaloa ($n=177$), estimated directly through the count of growth band counts in their vertebrae, showed that teleosts were the main preys in all age groups (%PSIRI = 62.8, 72 and 68.4 for 0–1, 1–2 and 2–3 years age groups, respectively). Nevertheless, the importance of crustaceans in the diet decreased with age (%PSIRI = 24.8, 15.3 and 4.9), whereas cephalopods increased in importance (%PSIRI = 12.5, 12.7 and 26). The gradual change of preferred prey with age might be related to incursions into oceanic waters. *Sphryna lewini* has been characterized by the traditional model of ontogenetic distribution proposed for sharks, in which juveniles reside in coastal areas from birth until they reach maturity, when they migrate to the adult's feeding grounds, as a strategy to avoid predation and have access to different sources of food (Klimley and Nelson, 1981; Knip et al., 2010). The overall trophic level of juvenile sharks (4.1) indicated that juveniles in this region are a tertiary predator in the ecosystem (Rentería-Bravo, 2016; Rentería-Bravo et al., 2019).

## 3.6 Gulf of Tehuantepec (GT)

The Gulf of Tehuantepec (GT) is characterized by intense oceanographic physical dynamics (Lavín and Marinone, 2003). Changes in salinity, temperature, and dissolved oxygen fluctuate according to wet and dry seasons, with two subsystems. The Oaxaqueño subsystem, with upwelling periods that lead to low temperatures, low dissolved oxygen, and high nutrient concentration, and the Chiapaneco subsystem, with strong influence of coastal lagoons and river discharges (Tapia-Garcia et al., 2007). The abiotic characteristics of the GT supports complex benthic, benthopelagic, and pelagic food webs, which strongly influence the distributional patterns of many shark species in the region.

The Silky Shark (*Carcharhinus falciformis*) is generally considered a circumtropical oceanic and coastal-pelagic species (Compagno, 1984). In the GT, adult sharks dominate the catches, (Torres-Rojas et al., 2013); however, Ronquillo (1999) mention that silky shark use the coastal waters of south-west Mexico as nursery grounds Thus, this region may serve as an important feeding area for this shark species. The trophic spectrum of *C. falciformis* was composed of two cephalopod species, one crustacean species, and eight fish species. Based on the index of relative importance (%IRI), *Dosidicus gigas* (34.0%), *Diodon hystrix* (21.7%), *Euthynnus lineatus* (17.6%) and *Isopistus remifer* (10.1%) were the most important components in the diet (Flores-Martinez et al., 2016). The prey consumed are mostly epipelagic-oceanic (e.g. *E. lineatus*) and mesopelagic oceanic (e.g. *D. gigas*) but also coastal species (e.g. *D. hystrix*), which indicate that this shark species prefers oceanic habitats where it consumes cephalopods but also migrates to the coastal zone where it consumes nearshore prey. Also Cabrera Chavez-Costa (2000) found that the crab *Portunus xanthusii affinis*, and the cephalopod *Gonatus* spp. were the most important prey for *C. falciformis*. Barranco (2008), on the other hand, found that the main prey of this shark were *Portunus xantusii*, cephalopods *Argonauta* spp., and the Black Skipjack *Euthynnus lineatus*. Using stable isotopes, Barranco (2008) found that silky shark fed on prey from oceanic waters ($\delta^{13}C$ −16.70 to −14.55‰). The trophic level of the silky shark was 4.34, which designates this species as a tertiary consumer.

The ontogenetic feeding behaviour of *C. falciformis* indicates differences in diet composition, suggesting trophic segregation. Cabrera Chavez-Costa et al. (2010) found differences in the main prey preferred by juveniles vs. adults, where juveniles fed on *D. gigas* and larger specimens consumed more *Scomber japonicus*. Segregation by gender and maturity stage is common in other Carcharhinids (Lowe et al., 1996; Torres-Rojas et al., 2013), and is probably associated with reducing food competition. The niche variation hypothesis (Giller, 1984) states that niche width tends to increase with intraspecific competition, either through an increase in the within-phenotype component of the niche (behavioural flexibility) or an increase in the between-phenotype component (differences among individuals). Flores-Martinez et al. (2016) report high niche width and low trophic overlap values observed for *C. falciformis*, which may reflect intraspecific competition where adult and juvenile males are found offshore but consume different main prey species (*Euthynus lineatus* for adult males, *D. gigas* for juvenile males) and juvenile females are found inshore where they feed on *Diodon hystrix*.

In contrast to *C. falciformis*, *S. lewini* is a coastal and semi-oceanic hammerhead shark (Torres-Rojas et al., 2013). The prey consumed by *S. lewini* (16 species) in the GT are both epipelagic (e.g. *C. orqueta*) and demersal (e.g. *Pomadasys panamensis*) (Flores-Martinez et al., 2016), which are characterized by being distributed in waters <100 m deep (coastal habits). The latter provides further evidence that *S. lewini* undertakes vertical migrations in the water column as reported in the GC (Klimley, 1987). The lack of differences in diet of juvenile *S. lewini* in the GT, suggests that males and females of the same size feed on similar prey species in the same area (Flores-Martinez et al., 2016). This pattern has been reported for the GC (Torres-Rojas et al., 2006, 2009, 2013), that is, juveniles feeding on small fish and crustaceans in benthic areas (Klimley, 1987) probably related with segregation by size to avoid cannibalism (Bush, 2003; Lowe et al., 1996). Cabrera Chavez-Costa (2000) found that *S. lewini* feeds on various crustaceans, the squid *Ancistrocheirus* spp., and on fishes of the families Carangidae, Cynoglossidae, Mullidae, Ophichtydae, Ophididae, and Serranidae. Whereas Aguilar-Castro (2010) found that the shrimp *Litopenaeus vannamei*, the squid *Mastigoteuthis dentata* and the fishes *Auxis* spp. and *Euthynnus lineatus* were the most important prey in the diet of *S. lewini*.

Using stable isotopes, García (2018) found that the highest $\delta^{15}N$ values in both sexes of *S. lewini* were present in the neonate stage (14.1‰ females and 13.8‰ males) and the lowest values in adults (12.9‰ females and 12.7‰ males), which indicates feeding segregation by size. The average values of $\delta^{13}C$ changed little between stages in both sexes. This value was higher in females (−14.8 to −11.97‰) than males (−14.0 to −12.2‰), in both stages they feed in coastal areas. The decreasing trends of $\delta^{15}N$ and $\delta^{13}C$ suggest progressive changes in the diet as ontogenetic development progresses, rather than a sudden change in diet at a given size.

A diet comparison between *S. lewini* and *C. falciformis* in the GT reflects low trophic overlap, likely as a result of spatial segregation (Flores-Martinez et al., 2016). Potier et al. (2007) found that the presence of epipelagic and mesopelagic prey species can be used to categorize predators as primarily shallow or deep feeders, respectively. Flores-Martinez et al. (2016) found that *C. falciformis* were predominantly epipelagic feeders preying on scombrid fishes (*E. lineatus*). Demersal prey, such as *P. panamensis* and *C. orqueta*, were more abundant in the *S. lewini* diet. Alejo-Plata et al. (2007) report high catch rates of *C. falciformis* with pelagic fishing gear, while *S. lewini* are more commonly caught with demersal fishing gears. Thus, although both shark species were captured in the GT, it is well known that *C. falciformis* benefit more from the pelagic food chain than *S. lewini* (Torres-Rojas et al., 2015).

Another coastal shark present in the GT is the Pacific Sharpnose Shark (*Rhizoprionodon longurio*), which is a common coastal species of the eastern tropical Pacific and is important in the artisanal fisheries in Mexico. Feeding studies on this species in Mexico have shown differences in some aspects of its food habits. For the GT, 37 prey species were identified. The IRI showed that the main prey of *R. longurio* in this area were benthic fish (e.g. congrid *Rynchoconger nitens* 59%) (Alderete, 2007). This was in contrast to the GC, where the prey species are both pelagic and benthic fishes (*Opisthopterus dovii*, 24%, family Ophichthidae, 18%, Bothidae, 17%) and cephalopods

(*Argonauta* spp., 15%, and *Loliolopsis diomedeae*, 12.3%). Based on isotopic composition the juveniles and adults feeding behaviours are similar, which means that all Sharpnose Sharks feed in the same areas (Conde, 2009). On the other hand, the same author found significant differences between the diets of juvenile males and adults in the GC, which coincides with other studies in the region (Alatorre-Ramirez et al., 2013). These differences might be because *R. longurio* of the two areas (GT and GC) do not migrate between them, which mean that they are separate populations or stocks. Another shark species occurring in the GT is *Nasolamia velox*, which feeds mainly on *Portunus xanthusiii affinis*, *Squilla biformis* and fishes of the family Ophichthydae (Cabrera Chavez-Costa, 2000).

## 4. Conservation status

Shark conservation and management proposals have been strengthened during the past decades as progress has been made on the huge task of assessing the conservation status of all sharks (around 524 species in total). This work has been undertaken by the International Union for the Conservation of Nature (IUCN) Species Survival Commission's Shark Specialist Group volunteer network, following the IUCN Red List Categories and Criteria (https://www.iucnredlist.org). The IUCN Red List of Threatened Species is widely recognized as the most comprehensive, scientifically based source of information on the global status of plant and animal species. Threatened species are listed as Critically Endangered (CR), Endangered (EN) or Vulnerable (VU). Species that are close to meeting the threatened thresholds are classified as Near Threatened (NT); while those evaluated as having a low risk of extinction are classified as Least Concern (LC). Finally, species that cannot be evaluated due to insufficient knowledge are classified as Data Deficient (DD). In this chapter, those new species (e.g. newly described species, new records) are listed as Not Evaluated (NE). Periodic species-specific evaluations mean that IUCN Red List assessments can be used as a tool for measuring and monitoring changes in the status of shark abundance and diversity over time (Cavanagh and Gibson, 2007).

A summary of the IUCN Red List category assigned to all shark species (60 species) from the Mexican Pacific is presented in Table 2. Thirty-one percent (18 species) of Mexican Pacific sharks are considered threatened (Critically Endangered, Endangered or Vulnerable). Of these, 5% (3 species) are Endangered and 25% (15 species) are Vulnerable (Fig. 6). The families with all species threatened are Alopiidae, Odontaspididae, and Rhincodontidae; while

**Table 2** Summary information on the IUCN Red List categories for all shark species recorded from the Pacific Coast of Mexico.

| SPECIES | Common name | Category | Source |
|---|---|---|---|
| *Alopias pelagicus* | Pelagic Thresher | VU | Reardon et al. (2009) |
| *Alopias superciliosus* | Bigeye Thresher | VU | Amorim et al. (2009) |
| *Alopias vulpinus* | Common Tresher Shark | VU | Goldman et al. (2009a) |
| *Apristurus brunneus* | Brown Catshark | DD | Huveneers et al. (2015a) |
| *Apristurus kampae* | Longnose Catshark | DD | Huveneers et al. (2015b) |
| *Apristurus nasutus* | Largenose Catshark | DD | Huveneers et al. (2004) |
| *Carcharhinus albimarginatus* | Silvertip Shark | VU | Espinoza et al. (2016) |
| *Carcharhinus altimus* | Bignose Shark | DD | Pillans et al. (2009) |
| *Carcharhinus brachyurus* | Copper Shark | NT | Duffy and Gordon (2003) |
| *Carcharhinus cerdale* | Pacific Smalltail Shark | NE | |
| *Carcharhinus falciformis* | Silky Shark | VU | Rigby et al. (2017) |
| *Carcharhinus galapagensis* | Galapagos Shark | NT | Bennett et al. (2003) |
| *Carcharhinus leucas* | Bull Shark | NT | Simpfendorfer and Burgess (2009) |
| *Carcharhinus limbatus* | Blacktip Shark | NT | Burgess and Branstetter (2009) |
| *Carcharhinus longimanus* | Oceanic Whitetip Shark | VU | Baum et al. (2015) |
| *Carcharhinus obscurus* | Dusky Shark | VU | Musick et al. (2009a) |
| *Carcharhinus porosus* | Smalltail Shark | DD | Lessa et al. (2006) |
| *Carcharhinus plumbeus* | Sandbar Shark | VU | Musick et al. (2009b) |
| *Carcharodon carcharias* | White Shark | VU | Fergusson et al. (2009) |
| *Cephaloscyllium ventriosum* | Swellshark | LC | Villavicencio-Garayzar et al. (2015) |

*Continued*

**Table 2** Summary information on the IUCN Red List categories for all shark species recorded from the Pacific Coast of Mexico.—cont'd

| SPECIES | Common name | Category | Source |
|---|---|---|---|
| *Cephalurus cephalus* | Lollipop Catshark | DD | Valenti (2009) |
| *Cetorhinus maximus* | Basking Shark | VU | Fowler (2009) |
| *Chlamydoselachus anguineus* | Lizard Shark | LC | Smart et al. (2016) |
| *Echinorhinus cookei* | Prickly Shark | DD | Finucci (2018) |
| *Galeocerdo cuvier* | Tiger Shark | NT | Simpfendorfer (2009) |
| *Galeorhinus galeus* | School Shark | VU | Walker et al. (2006) |
| *Galeus piperatus* | Peppered Catshark | LC | Heupel (2006) |
| *Ginglymostoma cirratum* | Nurse Shark | DD | Rosa et al. (2006) |
| *Ginglymostoma unami* | Pacific Nurse Shark | NE | |
| *Heterodontus francisci* | Horn Shark | DD | Carlisle (2015) |
| *Heterodontus mexicanus* | Mexican Hornshark | DD | Garayzar (2006) |
| *Hexanchus griseus* | Bluntnose Sixgill Shark | NT | Cook and Compagno (2005) |
| *Isistius brasiliensis* | Cookiecutter Shark | LC | Kyne (2018) |
| *Isurus oxyrinchus* | Shortfin Mako Shark | VU | Cailliet et al. (2009) |
| *Isurus paucus* | Longfin Mako | VU | Reardon et al. (2006) |
| *Lamna ditropis* | Salmon Shark | LC | Goldman et al. (2009b) |
| *Megachasma pelagios* | Megamouth Shark | LC | Simpfendorfer and Compagno (2015) |
| *Mustelus albipinnis* | White-Margin Fin Houndshark | DD | Cronin (2009) |
| *Mustelus californicus* | Gray Smooth-Hound | LC | Perez-Jimenez et al. (2015) |
| *Mustelus dorsalis* | Sharptooth Smooth-Hound | DD | Leandro (2004) |

**Table 2** Summary information on the IUCN Red List categories for all shark species recorded from the Pacific Coast of Mexico.—cont'd

| SPECIES | Common name | Category | Source |
|---|---|---|---|
| *Mustelus henlei* | Brown Smooth-Hound | LC | Perez-Jimenez et al. (2016a) |
| *Mustelus lunulatus* | Sicklefin Smooth-Hound | LC | Perez-Jimenez et al. (2016b) |
| *Nasolamia velox* | Whitenose Shark | DD | Ruiz et al. (2009) |
| *Negaprion brevirostris* | Lemon Shark | NT | Sundström (2015) |
| *Notorynchus cepedianus* | Broadnose Sevengill Shark | DD | Compagno (2009) |
| *Odontaspis ferox* | Smalltooth Sandtiger | VU | Graham et al. (2016) |
| *Parmaturus xaniurus* | Filetail Catshark | LC | Flammang et al. (2015) |
| *Prionace glauca* | Blue Shark | NT | Stevens (2009) |
| *Pseudocarcharias kamoharai* | Crocodile Shark | NT | Compagno and Musick (2009) |
| *Rhincodon typus* | Whale Shark | EN | Pierce and Norman (2016) |
| *Rhizoprionodon longurio* | Pacific Sharpnose Shark | DD | Smith et al. (2009b) |
| *Somniosus pacificus* | Pacific Sleeper Shark | DD | Ebert et al. (2009) |
| *Sphyrna corona* | Scalloped Bonnethead | NT | Mycock (2004) |
| *Sphyrna lewini* | Scalloped Hammerhead | EN | Baum et al. (2007) |
| *Sphyrna media* | Scoophead Shark | DD | Casper and Burgess (2006) |
| *Sphyrna mokarran* | Great Hammerhead | EN | Denham et al. (2007) |
| *Sphyrna tiburo* | Bonnethead Shark | LC | Cortés et al. (2016) |
| *Sphyrna zygaena* | Smooth Hammerhead | VU | Casper et al. (2005) |
| *Squalus suckleyi* | Spotted Spiny Dogfish | LC | Bigman et al. (2016) |
| *Squatina californica* | Pacific Angel shark | NT | Cailliet et al. (2016) |
| *Triaenodon obesus* | Whitetip Reef Shark | NT | Smale (2009) |
| *Triakis semifasciata* | Leopard Shark | LC | Carlisle et al. (2015) |

EN, endangered; VU, vulnerable; NT, near threatened; LC, least concern; DD, data deficient; NE, not evaluated.

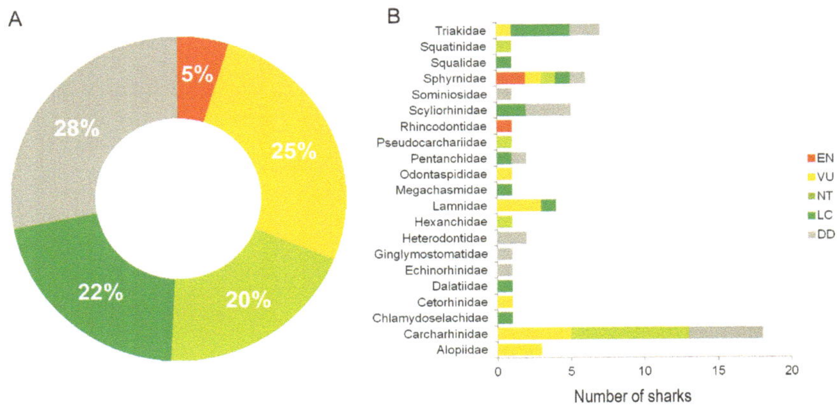


**Fig. 6** Percentage of shark species ($n = 61$) from the Pacific Coast of Mexico within each IUCN Red List Category (A), and number of sharks evaluated by families (B). IUCN Categories: EN, Endangered; VU, Vulnerable; NT, Near Threatened; LC, Least Concern; DD, Data Deficient. The species Not Evaluated are not included.

the families with the highest number of species threatened are Lamnidae (75%), Sphyrnidae (50%), and Squalidae (50%) (Fig. 6). These threatened species must be monitored particularly closely and appropriate management and recovery plans for their populations should be implemented. Additionally, more scientific research should be conducted to better understand these species' biology, ecology, threats, and conservation needs.

A further 20% (12 species) of all Mexican Pacific sharks are Near Threatened, which means that they are close to qualifying for the threatened category in the foreseeable future, and about 22% (13 species) are catalogued as Least Concern. The remaining sharks 17 (28%) are Data Deficient, meaning that there is not enough information to assess their status reliably. Additionally, two species of sharks, the Pacific Nurse Shark *Ginglymostoma unami* and the Pacific Smalltail Shark *Carcharhinus cerdale*, have not yet been assessed.

The proportion of shark species in each of the IUCN Red List categories varies along the Pacific Coast of Mexico (Fig. 7). The highest percentage of threatened species (43%) is observed in the Mexican Pacific Central, followed by Gulf of Tehuantepec (38%), while lower percentages are in the West coast of Baja California (32%) and the Gulf of California (30%).

The principal driver of Mexican Pacific sharks decline is overfishing and habitat degradation as has been observed in other areas (e.g. Dulvy et al., 2016). Sharks have historically been an important food resource in Mexico (Applegate et al., 1993). The first record of shark catches in Mexico goes

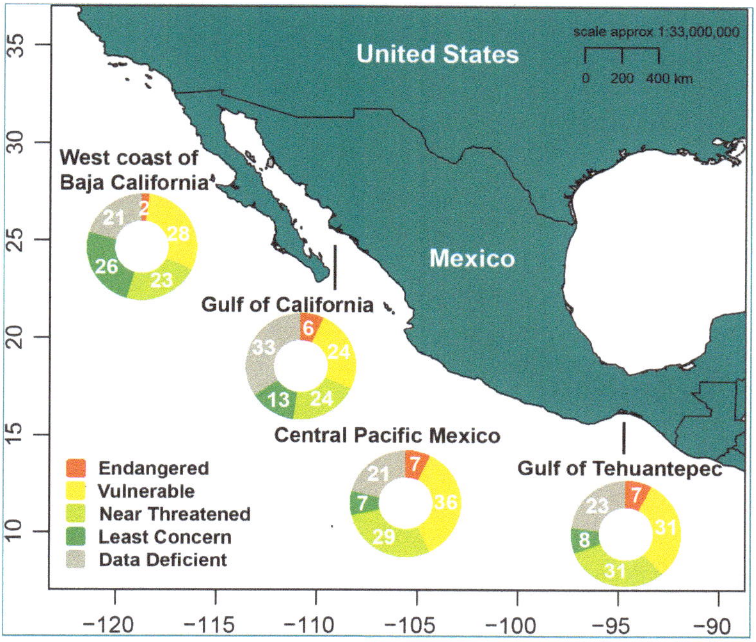


**Fig. 7** Percentage of shark species in each IUCN status category for four marine areas along the Pacific Coast of Mexico: Western coast of Baja California ($n=44$), Gulf of California ($n=44$), Central Pacific Mexico ($n=28$), and Gulf of Tehuantepec ($n=26$). The species Not Evaluated are not included.

backs to 1890, when shark fins were first exported to Asia from the Baja California Peninsula. Since that time shark fisheries have increased dramatically in their size and geographical extent. Additionally, sharks are an important constituent of the Mexican diet and up to 90% of the Mexican harvest is consumed domestically (Bonfil, 1997).

Although in Mexico there are few industrial fisheries for sharks, an important proportion of the landings are actually produced by artisanal fisheries like in other parts of the world, especially in developing countries (Bonfil, 2002; Humber et al., 2017; Ramírez-Amaro and Galván-Magaña, 2019).

The rapid expansion of shark artisanal fishery over the past two decades has greatly increased fishing pressure on shark populations, which constitute a high percentage of the total world catch (Selgrath et al., 2018). In addition, sharks caught as bycatch represent approximately half of the reported shark landings according to worldwide official statistics, which makes it difficult to accurately assess the impact of fishing on elasmobranch populations (Bonfil, 2002).

Management of a sustainable shark fishery in Mexico has been complicated by the lack of reliable fisheries data. For example, historical shark landings data are grouped into five broad categories: tiburón (sharks >1.5 m in length), cazón (sharks <1.5 m), angelito (angel sharks), manta (batoids), and guitarra (guitarfishes) (SAGARPA, 2009). This reporting scheme gives little information about species composition and ignores the distinction between adults of small shark species and juveniles of large sharks (Ramírez-Amaro and Galván-Magaña, 2019). Moreover, historical landings probably greatly underestimate the actual take due to a lack of adequate record keeping (Castillo-Geniz et al., 1998).

Various management measures have been implemented for Mexico's shark fisheries. The first step for management was taken by the Mexican National Fisheries and Aquaculture Institute (INAPESCA), which recommended a moratorium on issuing new shark-fishing permits beginning in 1993 (Castillo-Geniz et al., 1998), which was carried out in 1998 (Sosa-Nishizaki et al., 2008). This was followed by the development of a National Action Plan for the Conservation and Management of sharks, rays, and related species in Mexico in 2004 by the Comision Nacional de Agricultura y Pesca and INAPESCA (CONAPESCA-INP, 2004).

In 2007, a major piece of legislation was adopted called 'Responsible Fishery Sharks and Rays; Specification for Use'. It includes regulations that are specific to artisanal fisheries (NOM-059-ECOL-2001; NOM-029-PESC-2006; DOF, Diario Oficial de la Federacion, 2007). This law prohibits the capture of three sharks: the Whale Shark *Rhincodon typus*, White Shark *Carcharodon carcharias*, and Basking Shark *Cetorhinus maximus*, as well as specifies fishing zones and seasons, authorizes gears, and requires permit holders to report data (DOF, Diario Oficial de la Federacion, 2007). In addition, since 2012 the fishing of shark and rays has been closed annually between 1 May and 31 July (DOF, Diario Oficial de la Federacion, 2012), mainly to protect pregnant females of several elasmobranch species during their reproductive season, particularly those that approach to the coast like *S. lewini* making them highly susceptible to be caught.

In 2015, the governments of Canada, the United States, and Mexico initiated a collaborative project through the Commission for Environmental Cooperation (CEC) to strengthen the conservation and sustainable trade of 56 North America taxa that are included in Appendix II of the Convention on International Trade in Endangered Species of Wild Fauna and Flora (CITES). The project aims to provide guidance in the form of five action plans for reducing illegal and/or unsustainable harvest and trade; improving biological knowledge to allow science-based management decisions; and promoting traceability, species conservation, and livelihoods of stakeholders, throughout trade (CEC, 2017). Eight priority shark taxa were selected: *Carcharhinus longimanus*, *Carcharodon carcharias*, *Cetorhinus maximus*, *Lamna nasus*, *Rhincodon typus*, *Sphyrna lewini*, *Sphyrna mokarran*, and *Sphyrna zygaena*. All of these species are currently listed in Appendix II of CITES (CITES, 2016). Note that all these species, with an exception of *Lamna nasus*, are distributed along the Pacific coast of Mexico.

Feeding studies, since traditional stomach content analysis to stable isotopes analyses has provides insights into the trophic relationship among sharks and the ecosystems they inhabit (Cortés, 1999; Dicken et al., 2017). Knowledge of the diet and trophic ecology of sharks improve our understanding of the ecological consequences of predator or mesopredator. Therefore, these studies provided wealth information on the status of the ecosystems within different geographic regions. In this sense, more feeding studies are necessary for the implementation of effective management initiatives in Mexico. All these management measures will initiate the recovery of sharks, whose populations play an important role in marine ecosystems, in the Pacific coast of Mexico. These management measures will also help to develop true adaptive management in the area, making the fisheries compatible with the conservation of marine ecosystems.

## References

Abrams, P., 2000. The evolution of predator–prey interactions: theory and evidence. Annu. Rev. Ecol. Syst. 33, 79–105.

Aguilar-Castro, N.A., 2003. Ecología trófica de juveniles del tiburón martillo *Sphyrna lewini* (Griffith & Smith, 1834) en el Golfo de California. Instituto Politécnico Nacional, Centro Interdisciplinario de Ciencias Marinas. M.S. Thesis. 113 pp.

Aguilar-Castro, N.A., 2010. Análisis trófo-dinámico del tiburón Martillo *Sphyrna lewini* (Griffith and Smith 1834) en el Golfo de Tehuantepec. México. Universidad Autónoma de Baja California Sur. Ph d. Thesis. 64 pp.

Alatorre-Ramirez, V.G., Galvan-Magaña, F., Torres-Rojas, Y.E., 2013. Trophic habitat of the Pacific sharpnose shark, *Rhizoprionodon longurio*, in the Mexican Pacific. J. Mar. Biol. Assoc. U. K. 93, 2217–2224.

Alderete, M.J., 2007. Hábitos alimenticios del tiburón horma *Rhizoprionodon longurio* (Jordan y Gilbert 1882) en Salina Cruz. Oaxaca. Universidad Autónoma de Baja California Sur. Bachelor Thesis. 68 pp.

Alejo-Plata, M.C., Gomez-Marquez, J.L., Ramos, S., Herrera, E., 2007. Presencia de neonatos y juveniles del tiburón martillo *Sphyrna lewini* (Griffith & Smith, 1834) y del tiburón sedoso *Carcharhinus falciformis* (Muller & Henle, 1839) en la costa de Oaxaca, México. Rev. Biol. Mar. Oceanogr. 42, 403–413.

Amorim, A., Baum, J., Cailliet, G.M., Clò, S., Clarke, S.C., Fergusson, I., Gonzalez, M., Macias, D., Mancini, P., Mancusi, C., Myers, R., Reardon, M., Trejo, T.,

Vacchi, M., Valenti, S.V., 2009. Alopias superciliosus. In: The IUCN Red List of Threatened Species 2009, International Union for Conservation of Nature (IUCN), e.T161696A5482468.
Anislando-Tolentino, V., González-Medina, G., Ramos-Carrillo, S., 2008. Evaluacion respecto a los ataques de tiburón acontecidos en zonas aledañas a Zihuatanejo, Guerrero, durante abril y mayo de 2008. Universidad del Mar, Puerto Ángel, Oaxaca.
Applegate, S.P., Soltelo-Macias, F., Espinosa-Arrubarrena, L., 1993. An Overview of Mexican Shark Fisheries, With Suggestions for Shark Conservation in Mexico. vol. 115. NOAA Tech. Rep. NMFS, pp. 31–37.
Arriaga-Cabrera, L., Vázquez-Domínguez, E., González-Cano, J., Jiménez-Rosenberg, R., Muñoz-López, E., Aguilar-Sierra, V., 1998. Regiones marinas prioritarias de México. Comisión Nacional para el Conocimiento y uso de la Biodiversidad (CONABIO), México.
Barajas-Calderón, A.V., 2015. Hábitos alimentarios de juveniles del tiburón piloto *Carcharhinus falciformis* en la costa sur de Jalisco, México. Tesis de Licenciatura, Universidad de Guadalajara. 87 p.
Barajas-Calderón, A.V., 2018. Ecología trófica del tiburón piloto *Carcharhinus falciformis* (Müller & Henle, 1839) en la costa sur de Jalisco, México. Instituto Politécnico Nacional, Centro Interdisciplinario de Ciencias Marinas. M.S. Thesis 67 p.
Baro, I., 2016. Hábitos alimenticios del tiburón gato *Cephaloscyllium ventriosum* (Garman, 1880), en Bahía Tortugas, Baja California Sur, México. M.S. Thesis, Universitat de Barcelona.
Barranco, S.L.M., 2008. Hábitos alimenticios y nivel trófico del tiburón *Carcharhinus falciformis*, Mûller & Henle 1841 (Elasmobranchii: Carcharhinidae) en el Golfo de Tehuantepec, México por medio del análisis del contenido estomacal e isótopos estables de δ13C y δ15N. Universidad del Mar, Puerto Angel, Oaxaca. Master degree thesis. 86 pp.
Bascompte, J., Melián, C.J., Sala, E., 2005. Interaction strength combinations and the overfishing of a marine food web. Proc. Natl. Acad. Sci. U. S. A. 102, 5443–5447.
Baum, J., Clarke, S., Domingo, A., Ducrocq, M., Lamonaca, A.F., Gaibor, N., Graham, R., Jorgensen, S., Kotas, J.E., Medina, E., Martinez-Ortiz, J., Monzini Taccone di Sitizano, J., Morales, M.R., Navarro, S.S., Perez-Jimenez, J.C., Ruiz, C., Smith, W., Valenti, S.V., Vooren, C.M., 2007. *Sphyrna lewini*. In: The IUCN Red List of Threatened Species 2007, International Union for Conservation of Nature (IUCN), e.T39385A10190088.
Baum, J., Medina, E., Musick, J.A., Smale, M., 2015. *Carcharhinus longimanus*. In: The IUCN Red List of Threatened Species 2015, International Union for Conservation of Nature (IUCN), e.T39374A85699641.
Bennett, M.B., Gordon, I., Kyne, P.M., SSG Australia & Oceania Regional Workshop, 2003. March 2003 *Carcharhinus galapagensis*. The IUCN Red List of Threatened Species 2003. International Union for Conservation of Nature (IUCN), e.T41736A10550977.
Benson, J.F., Jorgensen, S.J., O'Sullivan, J.B., Winkler, C., White, C.F., Garcia-Rodriguez, E., Sosa-Nishizaki, O., Lowe, C.G., 2018. Juvenile survival, competing risks, and spatial variation in mortality risk of a marine apex predator. J. Appl. Ecol. 55 (6), 2888–2897.
Bigman, J.S., Ebert, D.A., Goldman, K.J., 2016. *Squalus suckleyi*. In: The IUCN red list of threatened species 2016, International Union for Conservation of Nature (IUCN), e.T195488A2382480.
Bizzarro, J.J., Smith, W.D., Jones, E.J., Cailliet, G.M., 2000. The artisanal elasmobranch fishery of Baja California Norte (Gulf of California, Mexico). In: American Elasmobranch Society, (Ed.), 80th Annual Meeting of the American Society of Ichthyologists and Herpetologists & 16th Annual Meeting of the American Elasmobranch Society, 14–20 June 2000. La Paz, Baja California Sur, México.

Bizzarro, J.J., Smith, W.D., Hueter, R.E., Villavicencio-Garayzar, C.J., 2009a. Activities and catch composition of artisanal elasmobranch fishing sites on the eastern coast of Baja California Sur, Mexico. Bull. South. Calif. Acad. Sci. 108, 137–151.

Bizzarro, J.J., Smith, W.D., Castillo-Geniz, J.L., Ocampo-Torres, A., Marquez-Farias, J.F., Hueter, R.E., 2009b. The seasonal importance of small coastal sharks and rays in the artisanal elasmobranch fishery in Sinaloa, Mexico. Pan Am. J. Aquat. Sci. 4, 513–531.

Bizzarro, J.J., Smith, W.D., Marquez-Farias, J.F., Tyminski, J., Hueter, R.E., 2009c. Temporal variation in the artisanal elasmobranch fishery of Sonora, Mexico. Fish. Res. 97, 103–117.

Bizzarro, J.J., Carlisle, A.B., Smith, W.D., Cortés, E., 2017. Diet composition and trophic ecology of Northeast Pacific Ocean sharks. In: Larson, S.E., Lowry, D. (Eds.), Northeast Pacific Shark Biology, Research and Conservation. In: Part A. Advances in Marine Biology, vol. 77. Academic Press, pp. 111–148.

Blanco-Parra, M.P., Galvan-Magaña, F., Marquez-Farias, F., Niño-Torres, C.A., 2012. Feeding ecology and trophic level of the banded guitarfish, *Zapterix exasperata*, inferred from stables isotopes and stomach contents analysis. Environ. Biol. Fishes 95 (1), 65–77.

Bonfil, R., 1997. Status of shark resources in the southern Gulf of Mexico and Caribbean: implications for management. Fish. Res. 29, 101–117.

Bonfil, R., 2002. Trends and patterns in world and Asian elasmobranch fisheries. In: Fowler, S.L., Reed, T.M., Dipper, F.A. (Eds.), Elasmobranch Biodiversity, Conservation and Management: Proceedings of the International Seminar and Workshop in Sabah. IUCN SSC Shark Specialist Group, Gland, Switzerland, Cambridge, UK, pp. 15–24.

Brusca, R., 2010. The Gulf of California: Biodiversity and Conservation. University of Arizona Press, Tucson, Arizona.

Brusca, R.C., Findley, L.T., Hastings, P.A., Hendrickx, M.E., Cosio, J.T., Van Der Heiden, A.M., 2005. Macrofaunal diversity in the Gulf of California. In: Cartron, J.L.E., Ceballos, G., Felger, R.S. (Eds.), Biodiversity, Ecosystems, and Conservation in Northern Mexico. Oxford University Press.

Burgess, H.G., Branstetter, S., 2009. *Carcharhinus limbatus*. In: The IUCN Red List of Threatened Species 2009, International Union for Conservation of Nature (IUCN), e.T3851A10124862.

Bush, A., 2003. Diet and diel feeding periodicity of juvenile scalloped hammerhead sharks *Sphyrna lewini*, in Kaneohe Bay, Oahu, Hawaii. Environ. Biol. Fishes 67, 1–11.

Bush, A., Holland, K.N., 2002. Food limitation in a nursery area: estimates of daily ration in juvenile scalloped hammerheads, *Sphyrna lewini* (Griffith and Smith, 1834) in Kaneohe Bay, Oahu, Hawaii. J. Exp. Mar. Biol. Ecol. 278, 157–178.

Cabrera Chavez-Costa, A.A., 2000. Determinación de los hábitos alimenticios durante las estaciones primavera y verano de *Carcharhinus falciformis*, *Sphyrna lewini* y *Nasolamia velox* (Carcharhiniformes:Carcharhinidae) a partir del análisis de su contenido estomacal en el Golfo de Tehuantepec, México. Universidad Nacional Autónoma de México. Bachelor thesis. 63 pp.

Cabrera Chávez-Costa, A.C., 2003. Hábitos alimenticios del tiburón piloto *Carcharhinus falciformis* (Bibron, 1839) en la costa occidental de Baja California Sur. M.S. Thesis, CICIMAR-IPN.

Cabrera Chavez-Costa, A.A., Galvan-Magaña, F., Escobar-Sanchez, O., 2010. Food habits of the silky shark *Carcharhinus falciformis* (Muller & Henle, 1839) off the western coast of Baja California Sur, Mexico. J. Appl. Ichthyol. 26, 499–503.

Cailliet, G.M., Musick, J.A., Simpfendorfer, C.A., Stevens, J.D., 2005. Ecology and life history characteristics of chondrichthyan fish. In: Fowler, S.L., Cavanagh, R.D., Camhi, M., Burgess, G.H., Fordham, S.V., Simpfendorfer, C.A., Musick, J.A. (Eds.), Sharks, Rays and Chimaeras: The Status of the Chondrichthyan Fishes. IUCN, Gland, Switzerland, pp. 12–18.

Cailliet, G.M., Cavanagh, R.D., Kulka, D.W., Stevens, J.D., Soldo, A., Clo, S., Macias, D., Baum, J., 2009. *Isurus oxyrinchus*. In: The IUCN Red List of Threatened Species 2009, International Union for Conservation of Nature (IUCN), e.T39341A10207466.
Cailliet, G.M., Chabot, C.L., Nehmens, M.C., Carlisle, A.B., 2016. *Squatina californica*. In: The IUCN Red List of Threatened Species 2016, International Union for Conservation of Nature (IUCN), e.T39328A80671059.
Camhi, M., Fowler, S., Musick, J., Bräutigam, A., Fordham, S., 1998. Sharks and their Relatives Ecology and Conservation. IUCN/SSC Shark Specialist Group. UICN, Cambridge, UK.
Carlisle, A.B., 2015. *Heterodontus francisci*. In: The IUCN Red List of Threatened Species 2015, International Union for Conservation of Nature (IUCN), e.T39333A80671300.
Carlisle, A.B., Smith, S.E., Launer, A.L., White, C.F., 2015. *Triakis semifasciata*. In: The IUCN Red List of Threatened Species 2015, International Union for Conservation of Nature (IUCN), e.T39363A80672743.
Cartamil, D., Santana-Morales, O., Escobedo-Olivera, M., Kacev, D., Castillo-Geniz, L., Graham, J.B., Rubin, R.D., Sosa-Nishizaki, O., 2011. The artisanal elasmobranch fishery of the Pacific coast of Baja California, Mexico. Fish. Res. 108, 393–403.
Casper, B.M., Burgess, G.H., 2006. *Sphyrna media*. In: The IUCN Red List of Threatened Species 2006, International Union for Conservation of Nature (IUCN), e.T60201A12317805.
Casper, B.M., Domingo, A., Gaibor, N., Heupel, M.R., Kotas, E., Lamonaca, A.F., Perez-Jimenez, J.C., Simpfendorfer, C., Smith, W.D., Stevens, J.D., Soldo, A., Vooren, C.M., 2005. *Sphyrna zygaena*. In: The IUCN Red List of Threatened Species 2005, International Union for Conservation of Nature (IUCN), e.T39388A10193797.
Castillo-Geniz, J.L., 1990. Contribución al conocimiento de la biología y pesquería del cazón bironche, *Rhizoprionodon longurio*, (Jordan y Gilbert, 1882) (Elasmobranchii, Carcharhinidae), del sur de Sinaloa, México. BSc thesis, Universidad Nacional Autónoma de México.
Castillo-Geniz, J.L., Márquez, F.J.F., Uribe, J., Bonfil, S.R., De Anda, D., Mena, R., Vélez, M.R., Mendizabal, O.D., 1996. La Pesquería de Tiburón en México. 365-415 En: Pesquerías Relevantes de México, XXX Aniversario del I.N.P., 1962-1992. Instituto Nacional de la Pesca, SEMARNAP.
Castillo-Geniz, J.L., Marquez-Farias, J.F., de la Cruz, M.C.R., Cortes, E., del Prado, A.C., 1998. The Mexican artisanal shark fishery in the Gulf of Mexico: towards a regulated fishery. Mar. Freshw. Res. 49, 611–620.
Castillo-Geniz, J.L., Soriano Velásquez, S.R., Sancho Vásquez, F., Ramírez Santiago, C.E., del Prado Vera, y.A.C., 2002. Elasmobranquios del Golfo de Tehuantepec, litoral chiapaneco. Universidad Autónoma de Baja California Sur. Informe final SNIB-CONABIO proyecto No. S123, México, D.F.
Castillo-Geniz, J.L., Godínez-Padilla, C.J., Ortega-Salgado, I., Ajás-Terriquez, H.A., Corro-Espinosa, D., Tovar-Ávila, J., Santana-Hernández, H., Valdez-Flores, J., Arellano-Torres, A., González-Cardenas, J.J., Meléndez-García, C., Rodríguez-Caballero, N.W., Morales-Pacheco, O., Hernández-Corona, J.L., Lizárraga-Rodríguez, A.A., Alatorre-Alba, A.J., Zea de la Cruz, H., Oviedo-Pérez, J.L., Martínez-Cruz, L.E., y González-Ocaranza, L., 2016. Capítulo 2, "La importancia pesquera de los tiburones incluidos en el Apéndice II de la CITES en aguas de México". Pp: 29-71. compiladores, In: Castillo-Géniz, J.L., Tovar-Ávila, y.J. (Eds.), Tiburones Mexicanos de Importancia pesquera en la CITES. Instituto Nacional de Pesca. 95 p.
Cavanagh, R.D., Gibson, C., 2007. Overview of the Conservation Status of Cartilaginous Fishes (Chondrichthyans) in the Mediterranean Sea. IUCN, Gland, Switzerland and Málaga.

CEC, 2017. Sustainable Trade in Sharks: Action Plan for North America. Commission for Environmental Cooperation, Montreal, Canada. 56 pp.

CITES, 2016. CITES Trade Database. CITES Secretariat, Geneva.

Clark, E., Nelson, D.R., 1997. Young whale sharks, *Rhincodon typus*, feeding on a copepod bloom near La Paz, Mexico. Environ. Biol. Fish. 50, 63–73.

Clarke, T.A., 1971. The ecology of the scalloped hammerhead shark, *Sphyrna lewini*, in Hawaii. Pac. Sci. 25, 133–144.

Colman, J., 1997. Whale Shark Interaction Management, with Particular Reference to Ningaloo Reef Marine Park, 1997–2007. Western Australia Wildlife Management Program.

Compagno, L.J.V., 1984. Sharks of the World: an annotated and illustrated catalogue of the shark species known to date. Part 2. Carcharhiniformes. In: Fischer, W., Nauen, C.E. (Eds.), FAO Species Catalogue. In: vol. 4. FAO, Rome, pp. 251–655. FAO Fisheries synopsis No. 125.

Compagno, L.J.V., 1990. Alternative life history styles of cartilaginous fishes in the time and space. Environ. Biol. Fishes 28, 33–75.

Compagno, L.J.V., 2001. Sharks of the World. An annotated and illustrated catalogue of shark species known to date. In: Bullhead, mackerel and carpet sharks (Heterodontiformes, Lamniformes and Orectolobiformes). vol. 2. FAO, RomeNo.1.

Compagno, L.J.V., 2009. *Notorynchus cepedianus*. In: The IUCN Red List of Threatened Species 2009, International Union for Conservation of Nature (IUCN), e.T39324A10200310.

Compagno, L.J.V., Musick, J.A., 2009. *Pseudocarcharias kamoharai*. In: The IUCN Red List of Threatened Species 2009, International Union for Conservation of Nature (IUCN), e.T39337A10204248.

CONAPESCA-INP, 2004. Plan de acción nacional para el manejo y conservación de tiburones, rayas y especies afines en México. In: Comisión Nacional de Acuacultura y Pesca e Instituto Nacional de la Pesca, Secretaria de Agricultura, Ganadería, Desarrollo Rural, Pesca y Alimentación. México, Mazatlán. 78 pp.

Conde, M.M., 2009. Ecología trófica del tiburón bironche, *Rhizoprionodon longurio* (Jordan & Gilbert, 1882), en dos áreas del pacífico mexicano. Master in Sciences. Thesis, CICIMAR-IPN La Paz, B.C.S. 94 pp.

Cook, S.F., Compagno, L.J.V., 2005. *Hexanchus griseus*. In: The IUCN Red List of Threatened Species 2005, International Union for Conservation of Nature (IUCN), e.T10030A3155348.

Cortés, E., 1999. Standardized diet composition and trophic levels of sharks. ICES J. Mar. Sci. 56, 707–717.

Cortés, E., 2000. Life history patterns and correlations in sharks. Rev. Fish. Sci. 8, 299–344.

Cortés, F.C., 2015. Hábitos alimenticios del tiburón dormilón cornudo *Heterodontus francisci* (girard 1855) en Bahía Tortugas. B.C.S. Bachelor thesis, Universidad de Baja California Sur.

Cortés, E., Lowry, D., Bethea, D., Lowe, C.G., 2016. *Sphyrna tiburo*. In: The IUCN Red List of Threatened Species 2016, International Union for Conservation of Nature (IUCN), e. T39387A2921446.

Cronin, E.S., 2009. *Mustelus albipinnis*. In: The IUCN Red List of Threatened Species 2009, International Union for Conservation of Nature (IUCN), e.T161451A5427107.

Denham, J., Stevens, J.D., Simpfendorfer, C., Heupel, M.R., Cliff, G., Morgan, A., Graham, R., Ducrocq, M., Dulvy, N.K., Seisay, M., Asber, M., Valenti, S.V., Litvinov, F., Martins, P., Lemine Ould Sidi, M., Tous, P., Bucal, D., 2007. *Sphyrna mokarran*. In: The IUCN Red List of Threatened Species 2007, International Union for Conservation of Nature (IUCN), e.T39386A10191938.

Dewar, H., Domeier, M., Nasby-Lucas, N., 2004. Insights into young of the year white shark, *Carcharodon carcharias*, behavior in the Southern California Bight. Environ. Biol. Fishes 70, 133–143.

Di Filippo, T.J., 2018. Ecología trófica del tiburón sulfin *Galeorhinus galeus* (Linnaeus, 1758) en Bahía Tortugas, Baja California Sur, México, mediante el análisis de contenido estomacal e isótopos estables. M.S. thesis, CICIMAR-IPN.

Dicken, M.L., Hussey, N.E., Christiansen, H.M., Smale, M.J., Nkabi, N., Cliff, G., Wintner, S.P., 2017. Diet and trophic ecology of the tiger shark (*Galeocerdo cuvier*) from South African waters. PLoS One 12 (6), e0177897.

DOF, Diario Oficial de la Federacion, 2007. Norma Oficial Mexicana NOM-029-PESC-2006, Pesca responsable de tiburones y rayas. Especificaciones para su aprovechamiento. Secretaria de Agricultura, Ganaderia, Desarrollo Rural, Pesca y Alimentacion (SAGARPA), Mexico.

DOF, Diario Oficial de la Federacion, 2012. Acuerdo por el que se modifica el Aviso por el que se da a conocer el establecimiento de epocas y zonas de veda para la pesca de diferentes especies de la fauna acuatica en aguas de jurisdiccion federal de los Estados Unidos Mexicanos, publicado el 16 de marzo de 1994 para establecer los periodos de veda de pulpo en el Sistema Arrecifal Veracruzano, jaiba en Sonora y Sinaloa, tiburones y rayas en el Oceano Pacifico y tiburones en el Golfo de Mexico. Secretaria de Agricultura, Ganaderia, Desarrollo Rural, Pesca y Alimentacion (SAGARPA), Mexico.

Domeier, M.L., 2012. A new life-history hypothesis for White sharks, *Carcharodon carcharias*, in the northeastern Pacific. In: Domeier, M.L. (Ed.), Global Perspectives on the Biology and Life History of the White Shark. Florida. CRC Press, Boca Raton, pp. 199–223.

Domeier, M., Nasby-Lucas, N., 2007. Annual re-sightings of photographically identified white sharks (*Carcharodon carcharias*) at an eastern Pacific aggregation site (Guadalupe Island, Mexico). Mar. Biol. 150, 977–984.

Duffy, C., Gordon, I., 2003. *Carcharhinus brachyurus*. The IUCN Red List of Threatened Species 2003, International Union for Conservation of Nature (IUCN), e.T41741A10551730.

Dulvy, N.K., Reynolds, J.D., 2002. Predicting extinction vulnerability in skates. Conserv. Biol. 16, 440–450.

Dulvy, N.K., Allen, D.J., Ralph, G.M., Walls, R.H.L., 2016. The Conservation Status of Sharks, Rays and Chimaeras in the Mediterranean Sea [Brochure]. IUCN, Malaga, Spain.

Ebert, D.A., 2003. Sharks, Rays and Chimeras of California. University of California Press. 285 pp.

Ebert, D.A., Mollet, H.F., Baldridge, A., Thomas, T., Forney, K.A., Ripley, W.E., 2004. Occurrence of the whale shark, *Rhincodon typus* Smith 1828, in California waters. Northwest. Nat. 85, 26–28.

Ebert, D.A., Goldman, K.J., Orlov, A.M., 2009. *Somniosus pacificus*. In: The IUCN Red List of Threatened Species 2009, International Union for Conservation of Nature (IUCN), e. T161403A5416294.

Ebert, D.A., Bigman, J.S., Lawson, J.M., 2017. Biodiversity, life history and conservation of Northeastern Pacific Chondrichthyans. In: Larson, S.E., Lowry, D. (Eds.), Northeast Pacific Shark Biology, Research and Conservation Part A. Academic Press, London.

Eckert, S.A., Stewart, B.S., 2001. Telemetry and satellite tracking of whale sharks, *Rhincodon typus*, in the sea of Cortez, Mexico, and the North Pacific Ocean. Environ. Biol. Fishes 60, 299–308.

Ehemann, N.R., González-González, L.V., Chollet-Villalpando, J.G., De la Cruz-Agüero, J., 2018. Updated checklist of the extant Chondrichthyes within the exclusive economic zone of Mexico. Zookeys 774, 17–39.

Ellis, J.R., Pawson, M.G., Shackley, S.E., 1996. The comparative feeding ecology of six species of shark and four species of ray (Elasmobranchii) in the north-east Atlantic. J. Mar. Biol. Assoc. U.K. 76, 89–106.

Ellis, J.K., Musick, J.A., 2007. Ontogenetic changes in the diet of the sandbar shark, *Carcharhinus plumbeus*, in lower Chesapeake Bay and Virginia (USA) coastal waters. Environ. Biol. Fishes 80, 51–60.

Escobar-Sanchez, O., Abitia-Cardenas, L.A., Galvan Magaña, F., 2006. Food habits of the Pacific angel shark *Squatina californica* in the southern gulf of California, Mexico. Cybium Int. J. Ichthyol. 30 (4), 91–97.

Escobar-Sanchez, O., Galvan-Magaña, F., Abitia-Cardenas, L.A., 2011. Trophic level and isotopic composition of δ13C and δ15N of Pacific angel shark, *Squatina californica* (Ayres, 1859), in the southern Gulf of California, Mexico. J. Fish. Aquat. Sci. 6 (2), 141–150.

Espinoza, G.S., 2012. Hábitos alimenticios del tiburón mamon *Mustelus henlei* (Gill 1863) en relación al sexo, la talla y la temporada en Punta Lobos, Baja California Sur, México. Bachelor thesis, Benemerita Universidad de Puebla.

Espinoza, M., González-Medina, E., Dulvy, N.K., Pillans, R.D., 2016. *Carcharhinus albimarginatus*. In: The IUCN Red List of Threatened Species 2016, International Union for Conservation of Nature (IUCN), e.T161526A68611084.

Fergusson, I., Compagno, L.J.V., Marks, M., 2009. *Carcharodon carcharias*. In: The IUCN Red List of Threatened Species 2009, International Union for Conservation of Nature (IUCN), e.T3855A10133872.

Ferretti, F., Worm, B., Britten, G., Heithaus, M.R., Lotze, H.K., 2010. Patterns and ecosystem consequences of shark declines in the ocean. Ecol. Lett. 13, 1055–1071.

Finucci, B., 2018. *Echinorhinus cookei*. In: The IUCN Red List of Threatened Species 2018, International Union for Conservation of Nature (IUCN), e.T41802A68622003.

Flammang, B.E., Cailliet, G.M., Ebert, D.A., 2015. *Parmaturus xaniurus*. In: The IUCN Red List of Threatened Species 2015, International Union for Conservation of Nature (IUCN), e.T60231A80671960.

Flores-Martinez, I.A., Torres-Rojas, Y.E., Galvan-Magaña, F., Ramos-Miranda, J., 2016. Diet comparison between silky sharks (*Carcharhinus falciformis*) and scalloped hammerhead sharks (*Sphyrna lewini*) off the south-west coast of Mexico. J. Mar. Biol. Assoc. U. K. 97, 1–9.

Fowler, S.L., 2009. *Cetorhinus maximus*. In: The IUCN Red List of Threatened Species 2009International Union for Conservation of Nature (IUCN), e.T4292A10763893.

Gallo-Reynoso, J.P., 1994. Factors affecting the population status of Guadalupe fur seal, *Arctocephalus townsendi* (Merriam, 1897), at Isla de Guadalupe, Baja California, México. Ph.D. ThesisUniversity of California, Santa Cruz.

Gallo-Reynoso, J.P., Figueroa-Carranza, A.L., Le Boeuf, B.J., 2008. Foraging behavior of lactating Guadalupe fur seal females. In: Lorenzo, C., Espinoza, E., Ortega, J. (Eds.), Avances en el estudio de los mamíferos de México. Asociación Mexicana de Mastozoología, A.C., Mexico City, pp. 595–614

Galvan-Magaña, F., Nienhuis, H., Klimley, A.P., 1989. Seasonal abundance and feeding habits of sharks of the lower gulf of California, Mexico. Calif. Fish Game 75 (2), 74–84.

Galván-Magaña, F., Hoyos-Padilla, E.M., Navarro-Serment, C.J., Márquez-Farías, F., 2010. Records of white shark, *Carcharodon carcharias*, in the Gulf of California, Mexico. Mar. Biol. Rec. 3, e111.

Garayzar, C.V., 2006. *Heterodontus mexicanus*. In: The IUCN Red List of Threatened Species 2006, International Union for Conservation of Nature (IUCN), e.T60235A12331644.

García, F.F., 2018. Cambios tróficos durante el desarrollo ontogénico del tiburón martillo *Sphyrna lewini* (Griffith y Smith 1834) en el Golfo de Tehuantepec, México. CICIMAR-IPN, La Paz, México, Master degree thesis. 54 pp.

Giller, P.S., 1984. Community Structure and the Niche. Outline Studies in Ecology. Chapman & Hall, London.

Goldman, K., Kohin, S., Cailliet, G.M., Musick, J.A., 2009a. *Lamna ditropis*. In: The IUCN Red List of Threatened Species 2009a, International Union for Conservation of Nature (IUCN), e.T39342A10210228.

Goldman, K.J., Baum, J., Cailliet, G.M., Cortes, E., Kohin, S., Macias, D., Megalofonou, P., Perez, M., Soldo, A., Trejo, T., 2009b. *Alopias vulpinus*. In: The IUCN Red List of Threatened Species 2009b, International Union for Conservation of Nature (IUCN), e.T39339A10205317.

Graham, K.J., Pollard, D.A., Gordon, I., Williams, S., Flaherty, A.A., Fergusson, I., Dicken, M., 2016. *Odontaspis ferox* (errata version published in 2016). In: The IUCN Red List of Threatened Species 2016, International Union for Conservation of Nature (IUCN), e.T41876A103433002.

Grubbs, R.D., 2010. Ontogenetic shifts in movements and habitat use. Sharks and their relatives II: Biodiversity, adaptive physiology, and conservation. CRC Press, Boca Raton, FL, pp. 319–350.

Hacohen-Domené, A., Galván-Magaña, F., Ketchum, J.T., 2006. Abundance of whale shark (*Rhincodon typus*) preferred prey in the southern gulf of California, Mexico. Cybium 30, 99–102.

Harley, C.D., 2011. Climate change, keystone predation, and biodiversity loss. Science 334, 1124.

Hendrickx, M.E., Salgado-Barragán, J., 1991. Los Estomatópodos (Crustacea: Hoplocarida) del Pacífico Mexicano. vol. 10. Publicaciones especiales del Instituto de Ciencias del Mar y Limnología. UNAM. 200 p.

Hernández-Aguilar, S.B., Escobar-Sánchez, O., Galván-Magaña, F., Abitia-Cárdenas, L.A., 2016. Trophic ecology of the blue shark *Prionace glauca* (Linnaeus 1758) based on stable isotopes (δ13C and δ15N) and stomach contents. J. Mar. Biol. Assoc. U. K. 96, 1403–1410. https://doi.org/10.1017/S0025315415001393.

Hernández-Corona, J.L., 2011. Hábitos alimenticios de *Sphyrna lewini* (Griffith y Smith, 1834) en Nayarit, México e importancia de la estimación de la talla de las presas para la cuantificación de la dieta. Tesis de licenciatura. Instituto Tecnológico de Bahía de Banderas. 79p.

Heupel, M.R., 2006. *Galeus piperatus*. In: The IUCN Red List of Threatened Species 2006, International Union for Conservation of Nature (IUCN), e.T60228A12329390.

Heyman, W.D., Graham, R.T., Kjerfve, B., Johannes, R.E., 2001. Whale sharks *Rhincodon typus* aggregate to feed on fish spawn in Belize. Mar. Ecol. Prog. Ser. 215, 275–282.

Hoyos-Padilla, E.M., 2009. Movement Patterns of the White Shark (*Carcharodon carcharias*) at Guadalupe Island, Mexico. Ph. D. Thesis, Centro Interdisciplinario de Ciencias Marinas, Instituto Politécnico Nacional, La Paz, B.C.S., Mexico.

Hoyos-Padilla, E.M., Ketchum, J.T., Klimley, A.P., Galván-Magaña, F., 2014. Ontogenetic migration of a female scalloped hammerhead shark in the Gulf of California. Anim. Biotelem. 2 (17), 1–9.

Hoyos-Padilla, E.M., Klimley, A.P., Galván-Magaña, F., Antoniou, A., 2016. Contrasts in the movements and habitat use of juvenile and adult white sharks (*Carcharodon carcharias*) at Guadalupe Island, Mexico. Anim. Biotelem. 4 (1), 14.

Humber, F., Adriamahaino, E.T., Beriziny, T., Botosoamananto, R., Godley, B.J., Gough, C., Pedron, S., Ramahery, V., Broderick, A.C., 2017. Assessing the small-scale shark fishery of Madagascar through community-based monitoring and knowledge. Fish. Res. 186, 131–143.

Huveneers, C., Duffy, C., Acuña, E., 2004. *Apristurus nasutus*. In: The IUCN Red List of Threatened Species 2004, International Union for Conservation of Nature (IUCN), e.T44573A10921377.

Huveneers, C., Duffy, C.A.J., Cordova, J., Ebert, D.A., 2015a. *Apristurus kampae*. In: The IUCN Red List of Threatened Species 2015, International Union for Conservation of Nature (IUCN), e.T44215A80671609.

Huveneers, C., Duffy, C.A.J., Cordova, J., Ebert, D.A., 2015b. *Apristurus brunneus*. In: The IUCN Red List of Threatened Species 2015, International Union for Conservation of Nature (IUCN), e.T44209A80671448.

Jaime-Rivera, M., Caraveo-Patiño, J., Hoyos-Padilla, M., Galván-Magaña, F., 2014. Feeding and migration habits of white shark Carcharodon carcharias (Lamniformes: Lamnidae) from Isla Guadalupe inferred by analysis of stable isotopes δ15N and δ13C. Int. J. Trop. Biol. 62 (2), 637–647.

Jiménez-Illescas, A.R., Obeso-Nieblas, M., Salas-León, D.A., 1997. Oceanografía física de la Bahía de La Paz, B.C.S. In: Urbán, R.J., Ramírez, M. (Eds.), La Bahía de La Paz, investigación y conservación. UABCS-CICIMAR-SCRIPPS, La Paz, pp. 31–41.

Kabasakal, H., Özgür-Gedikoğ, L.S., 2008. Two new-born great white sharks, *Carcharodon carcharias* (Linnaeus, 1758) (Lamniformes; Lamnidae) from Turkish waters of the North Aegean Sea. Acta Adriat. 49, 125–135.

Kessler, W.S., 2006. The circulation of the eastern tropical Pacific: a review. Prog. Oceanogr. 69, 181–217.

Ketchum, J.T., 2003. Distribución espacio-temporal y ecología alimentaria del tiburón ballena (*Rhincodon typus*) en la Bahía de La Paz y zonas adyacentes en el suroeste del Golfo de California. MSc. thesis. CICIMAR, La Paz. 91p.

Ketchum, J.T., Lozano, A.R., 2000. Observations on the seasonal occurrences of whale sharks (*Rhincodon typus*) in the southern Gulf of California. In: AES Symposium, Natural History of the Whale Shark. La Paz, Mexico.

Ketchum, J.T., Galván-Magaña, F., Klimley, A.P., 2013. Segregation and foraging ecology of whale sharks (*Rhincodon typus*) in the southwestern Gulf of California. Environ. Biol. Fishes 96 (6), 779–795.

Klimley, A.P., 1985a. Schooling in the large predator, Sphyrna lewini, a species with low risk of predation: a non-egalitarian state. Zeitschrift für Tierpsychologie (Ethology) 70, 297–319.

Klimley, A.P., 1985b. The area distribution and autoecology of the white shark, *Carcharodon carcharias*, off the west coast of North America. Mem. South. Calif. Acad. Sci. 9, 15–40.

Klimley, A.P., 1987. The determinants of sexual segregation in the scalloped hammerhead, *Sphyrna lewini*. Environ. Biol. Fishes 18, 27–40.

Klimley, A.P., 1993. Highly directional swimming by scalloped hammerhead sharks, *Sphyrna lewini*, and subsurface irradiance, temperature, bathymetry, and geomagnetic field. Mar. Biol. 117, 1–22.

Klimley, A.P., 1994. The predatory behavior of the white shark. Am. Sci. 82, 122–133.

Klimley, A.P., Butler, S.B., 1988. Immigration and emigration of a pelagic fish assemblage to seamounts in the Gulf of California related to water mass movements using satellite imagery. Mar. Ecol. Prog. Ser. 49, 11–20.

Klimley, A.P., Nelson, D.R., 1981. Schooling of scalloped hammerhead, *Sphyrna lewini*, in the Gulf of California. Fish. Bull. 79, 356–360.

Klimley, A.P., Nelson, D.R., 1984. Diel movement patterns of the scalloped hammerhead shark (*Sphyrna lewini*) in relation to El Bajo Espiritu Santo: a refuging central-position social system. Behav. Ecol. Sociobiol. 15, 45–54.

Knip, D.M., Heupel, M.R., Simpfendorfer, C.A., 2010. Sharks in nearshore environments: models, importance and consequences. Mar. Ecol. Prog. Ser. 402, 1–11.

Kyne, P.M., 2018. *Isistius brasiliensis*. In: The IUCN Red List of Threatened Species 2018, International Union for Conservation of Nature (IUCN), e.T41830A2956761.

Last, P.R., Stevens, J.D, 1994. Sharks and Rays of Australia. CSIRO, p. 513, Australia.

Lavín, M.F., Marinone, S.G., 2003. An overview of the physical oceanography of the Gulf of California. In: Velasco-Fuentes, O.U., Sheinbaum, J., Ochoa, J. (Eds.), Nonlinear Processes in Geophysical Fluid Dynamics. Kluwer Academic Publishers, The Netherlands, pp. 173–204.

Leandro, L., 2004. *Mustelus dorsalis*. In: The IUCN Red List of Threatened Species 2004, International Union for Conservation of Nature (IUCN), e.T44580A10908119.

Lessa, R., Almeida, Z., Santana, F.M., Siu, S., Perez, M., 2006. *Carcharhinus porosus*. In: The IUCN Red List of Threatened Species 2006, International Union for Conservation of Nature (IUCN), e.T60220A12324372.

Lowe, C.G., Wetherbee, B.M., Crow, G.L., Tester, A.L., 1996. Ontogenetic dietary shifts and feeding behaviour of the tiger shark, *Galeocerdo cuvier*, in Hawaiian waters. Environ. Biol. Fishes 47, 203–211.

Lucifora, L.O., García, V.B., Menni, R.C., Worm, B., 2012. Spatial patterns in the diversity of sharks, rays, and chimaeras (Chondrichthyes) in the SW Atlantic. Biodivers. Conserv. 21, 407–419.

Malpica-Cruz, M., Herzka, S.H., Sosa-Nishizaki, O., Escobedo-Olvera, M.A., 2013. Tissue-specific stable isotope ratios of shortfin mako (*Isurus oxyrinchus*) and white (*Carcharodon carcharias*) sharks as indicators of size-based differences in foraging habitat and trophic level. Fish. Oceanogr. 22, 551–553.

Markaida, U., Sosa-Nishisaki, O., 2010. Food and feeding habits of the blue shark Prionace glauca caught off Ensenada, Baja California, Mexico, with a review on its feeding. J. Mar. Biol. Assoc. UK 90 (5), 977–994.

Marquez-Farias, J.F., Corro-Espinosa, D., Castillo-Geniz, J.L., 2004. Observations on the iology of the Pacific sharpnose shark, *Rhizoprionodon longurio* (Jordan & Gilbert, 1882), captured in southern Sinaloa, México. J. Northwest Atlant. Fish. Sci. 35, 107–114.

Martínez, A.J.C., 2018. Ecología trófica del cazón mamon *Mustelus lunulatus* (Jordan & Gilbert, 1882) en la costa occidental de Baja California Sur, México. M.S. thesis, CICIMAR-IPN.

Maya-Meneses, C.I., Torres-Rojas, Y.E., Galván-Magaña, F., Aguiñiga-García, S., Trasviña-Carrillo, L.D., 2016. Trophic overlap between blue sharks (*Prionace glauca*) and shortfin mako (*Isurus oxyrinchus*): trophic linkages of two shark species in the Pacific Ocean food web. Food Webs 7 (2016), 13–19.

Méndez Da Silveira, E.D., 2015. Composición isotópica de Carbono y Nitrógeno (δ 13C y δ 15N) en tejidos de tiburones de la costa occidental de Baja California Sur. M.S. thesis, CICIMAR-IPN, México.

Mendez-Loeza, I., 2004. Hábitos alimentarios de los cazones *Mustelus californicus* y *Mustelus lunulatus* (triakidae) en el Alto Golfo de California. Bachelor thesis, UABC. 65 pp.

Molina-Cruz, A., Martinez-López, M., 1994. Oceanography of the Gulf of Tehuantepec, Mexico, indicted by Radiolaria remains. Paleogeogr. Paleoclimatol. Paleoecol. 110, 179–195.

Morales-Bojórquez, E., Hernández-Herrera, A., Nevárez-Martínez, M.O., Díaz-Uribe, J.G., 2012. Population size of the jumbo squid *Dosidicus gigas* in the central Gulf of California, Mexico, based on mark recapture data. Aquat. Biol. 15, 27–34.

Musick, J.A., Stevens, J.D., Baum, J.K., Bradai, M., Clò, S., Fergusson, I., Grubbs, R.D., Soldo, A., Vacchi, M., Vooren, C.M., 2009a. *Carcharhinus plumbeus*. In: The IUCN Red List of Threatened Species 2009, International Union for Conservation of Nature (IUCN), e.T3853A10130397.

Musick, J.A., Grubbs, R.D., Baum, J., Cortes, E., 2009b. *Carcharhinus obscurus*. In: The IUCN Red List of Threatened Species 2009, International Union for Conservation of Nature (IUCN), e.T3852A10127245.

Mycock, S.G., 2004. *Sphyrna corona*. In: The IUCN Red List of Threatened Species 2004, International Union for Conservation of Nature (IUCN), e.T44591A10910637.

Myers, R.A., Baum, J.K., Shepherd, T., Powers, S.P., Peterson, C.H., 2007. Cascading effects of the loss of apex predatory sharks from a coastal ocean. Science 315, 1846–1850.

Nelson, J.D., Eckert, S.A., 2007. Foraging ecology of whale sharks (*Rhincodon typus*) within Bahia de Los Angeles, Baja California Norte, Mexico. Fish. Res. 84, 47–64.

Obeso-Nieblas, M., Gaviño-Rodríguez, J.H., Jiménez-Illescas, A.R., Shirasago-Germán, B., 2002. Simulación numérica de la circulación por marea y viento del noroeste y sur en la Bahía de La Paz, B.C.S. Oceánides 11, 1–2.

Ochoa Diaz, M.R., 2009. Espectro trófico del tiburón martillo Sphyrna zygaena (Linnaeus, 1758) en Baja California Sur: aplicación de δ13C Y δ15N. M.S. Thesis, Instituto

Politécnico Nacional. Centro Interdisciplinario de Ciencias Marinas, La Paz, B.C.S., México. 82 p.

Oñate-González, E.C., Sosa-Nishizaki, O., Herzka, S.Z., Lowe, C.G., Lyons, K., Santana-Morales, O., Sepulveda, C., Guerrero-Avila, C., Garcia-Rodriguez, E., Ssullivan, O., J. B., 2017. Importance of Bahia Sebastian Vizcaino as a nursery area for white sharks (*Carcharodon carcharias*) in the northeastern Pacific: a fishery dependent analysis. Fish. Res. 188, 125–137.

Palomares-García, J.R., 1996. Estructura espacial y variación estacional de los copépodos en la Ensenada de La Paz. Oceánides 11, 29–43.

Perez-Jimenez, J., Vasquez, V.E., Chabot, C.L., Ebert, D.A., 2015. *Mustelus californicus*. In: The IUCN Red List of Threatened Species 2015, International Union for Conservation of Nature (IUCN), e.T161334A80672080.

Perez-Jimenez, J., Carlisle, A.B., Chabot, C.L., Vasquez, V.E., Ebert, D.A., 2016a. *Mustelus henlei*. In: The IUCN Red List of Threatened Species 2016, International Union for Conservation of Nature (IUCN), e.T161648A80672263.

Perez-Jimenez, J., White, C.F., Ruiz, C., Carlisle, A.B., Lowe, C.G., 2016b. *Mustelus lunulatus*. In: The IUCN Red List of Threatened Species 2016, International Union for Conservation of Nature (IUCN), e.T161640A80672480.

Pierce, S.J., Norman, B., 2016. *Rhincodon typus*. In: The IUCN Red List of Threatened Species 2016, International Union for Conservation of Nature (IUCN), e.T19488A2365291.

Pillans, R., Amorim, A., Mancini, P., Gonzalez, M., Anderson, C., 2009. *Carcharhinus altimus*. In: The IUCN Red List of Threatened Species 2009, International Union for Conservation of Nature (IUCN), e.T161564A5452406.

Pistevos, J.C.A., Nagelkerken, I., Rossi, T., Olmos, M., Connell, S.D., 2015. Ocean acidification and global warming impair shark hunting behaviour and growth. Sci. Rep. 5, 16293.

Polo, S.C.J., 2013. Ontogenia alimentaria del tiburón azul *Prionace glauca* (Linnaeus, 1758) a partir de análisis isotópicos en tejidos diferentes. Ph d. thesis, Instituto de Ciencias del Mar y Limnología. UNAM.

Polo-Silva, C., Galván-Magaña, F., Delgado-Huertas, A., 2012. Trophic inferences of blue shark (*Prionace glauca*) in the Mexican Pacific from stable isotope analysis in teeth. Rapid Commun. Mass Spectrom. 26, 1631–1638.

Potier, M., Marsac, F., Cherel, Y., Lucas, V., Sabatie, R., Maury, O., Menard, F., 2007. Forage fauna in the diet of three large pelagic fishes (lancetfish, swordfish and yellowfin tuna) in the western equatorial Indian Ocean. Fish. Res. 83, 60–72.

Prugh, L., Stoner, C., Epps, C., Beam, W., Ripple, W., Laliberte, A., Brashares, J., 2009. The rise of the mesopredator. Bioscience 59, 779–791.

Ramírez-Amaro, S., Galván-Magaña, F., 2019. Effect of gillnet selectivity on elasmobranchs off the northwestern coast of Mexico. Ocean Coast. Manag. 172, 105–116.

Ramirez-Amaro, S.R., Cartamil, D., Galvan-Magana, F., Gonzalez-Barba, G., Graham, J.B., Carrera-Fernandez, M., Escobar-Sanchez, O., Sosa-Nishizaki, O., Rochin-Alamillo, A., 2013. The artisanal elasmobranch fishery of the Pacific coast of Baja California Sur, Mexico, management implications. Sci. Mar. 77, 473–487.

Ramírez-Macías, D., Meekan, M., de la Parra-Venegas, R., Remolina-Suárez, F., Trigo-Mendoza, M., Vázquez-Juárez, R., 2012. Patterns in composition, abundance and scarring of whale sharks *Rhincodon typus* near Holbox Island, Mexico. J. Fish Biol. 80, 1401–1416.

Ramírez-Macías, D., Queiroz, N., Pierce, S.J., Humphries, N.E., Sims, D.W., Brunnschweiler, J.M., 2017. Oceanic adults, coastal juveniles: tracking the habitat use of whale sharks off the Pacific coast of Mexico. PeerJ 5, e3271.

Reardon, M.B., Gerber, L., Cavanagh, R.D., 2006. *Isurus paucus*. In: The IUCN Red List of Threatened Species 2006, International Union for Conservation of Nature (IUCN), e.T60225A12328101.

Reardon, M., Marquez, F., Trejo, T., Clarke, S.C., 2009. *Alopias pelagicus*. In: The IUCN Red List of Threatened Species 2009, International Union for Conservation of Nature (IUCN), e.T161597A5460720.

Rentería-Bravo, J., 2016. Estimación del tiempo de residencia de neonatos y juveniles de *Sphyrna lewini* (Griffith y Smith 1834) en áreas de crianza en la entrada al Golfo de California, con base en el análisis de sus contenidos estomacales y edad. Bachelor Thesis, Instituto Tecnológico de Bahía Banderas.

Rentería-Bravo, J., Tovar-Ávila, J., Chávez-Arrequín, D.A., Hernández-Corona, J.L., 2019. Variation in the Diet of Juvenile Scalloped Hammerhead Shark (*Sphyrna lewini*) as Evidence of their Migratory Habits at the Entrance to the Gulf of California (Nayarit and Sinaloa). Primer Congreso Latinoamericano de Tiburones, Rayas y Quimeras y VIII Simposium Nacional de Tiburones y Rayas. Playa Del Carmen, México.

Rigby, C.L., Sherman, C.S., Chin, A., Simpfendorfer, C., 2017. *Carcharhinus falciformis*. In: The IUCN Red List of Threatened Species 2017, International Union for Conservation of Nature (IUCN), e.T39370A117721799.

Rivera-Arriaga, E., Villalobos, G., 2001. The coast of Mexico: approaches for its management. Ocean Coast. Manag. 44, 729–756.

Rodríguez, C.J.M., 2016. Amplitud, traslape y posición trófica de *Heterodontus francisci* (Girard, 1854) capturado en Punta Eugenia, Baja California Sur, México: aplicación de δ13C y δ15N. Bachelor thesis, Universidad del Mar, Oaxaca, México.

Romero-Caicedo, A.F., Galván-Magaña, F., Hernández-Herrera, A., Carrera-Fernández, M., 2016. Reproductive parameters of the Pacific angel shark *Squatina californica* (Selachii: Squatinidae). J. Fish Biol. 88, 1430–1440. https://doi.org/10.1111/jfb.12920.

Ronquillo, K., 1999. Aspectos biológicos y pesqueros del tiburón aleta de carton *Carcharhinus falciformis* (Bibron, 1839) de las aguas del Golfo de Tehuantepec, Chiapas, México. Bachelor thesis, UNAM, México.

Rosa, R.S., Castro, A.L.F., Furtado, M., Monzini, J., Grubbs, R.D., 2006. *Ginglymostoma cirratum*. In: The IUCN Red List of Threatened Species 2006, International Union for Conservation of Nature (IUCN), e.T60223A12325895.

Ruiz, C., Arauz, R., Pérez-Jiménez, J.C., Castillo-Geniz, J.L., Soriano-Velásquez, S., 2009. *Nasolamia velox*. In: The IUCN Red List of Threatened Species 2009, International Union for Conservation of Nature (IUCN), e.T161355A5405297.

SAGARPA, 2009. Avisos de arribo de pesca de elasmobranquios. Subdelegación de Pesca en Baja California. Ensenada, B.C.

Saldaña-Ruiz, L.E., Sosa-Nishizaki, O., Cartamil, D., 2017. Historical reconstruction of gulf of California shark fishery landings and species composition, 1939-2014, in a data-poor fishery context. Fish. Res. 195, 116–129.

Salinas-de-León, P., Hoyos-Padilla, E.M., Pochet, E., 2017. First observation on the mating behaviour of the endangered scalloped hammerhead shark *Sphyrna lewini* in the Tropical Eastern Pacific. Environ. Biol. Fishes 100, 1603–1608. https://doi.org/10.1007/s10641-017-0668-0.

Santana-Morales, O., Sosa-Nishizaki, O., Escobedo-Olvera, M.A., Oñate-González, E.C., O'Sullivan, J.B., Cartamil, D., 2012. Incidental catch and ecological observations of juvenile white sharks, *Carcharodon carcharias*, in western Baja California, Mexico. In: Domeier, M.L. (Ed.), Global Perspectives on the Biology and Life History of the White Shark. CRC Press, Boca Raton, Florida, pp. 187–198.

Segura, Z.J.C., Abitia, C.L.A., Galván, M.F., 1997. Observaciones sobre la alimentación del tiburón *Heterodontus francisci* Girard, 1854 (Chondrichthyes:Heterodontidae) en Laguna San Ignacio, B.C.S., México. (In English). Cienc. Mar. 23 (1), 111–128.

Selgrath, J.C., Gergel, S.E., Vicent, A., 2018. Incorporating spatial dynamics greatly increases estimates of long-term fishing effort: a participatory mapping approach. ICES J. Mar. Sci. 75, 201–220.

Signoret, M., Santoyo, H., 1980. Aspectos ecológicos del plancton en la Bahía de La Paz, Baja California Sur. An. Cent. Cienc. Mar. Limnol. 7, 217–248.

Simpfendorfer, C., 2009. *Galeocerdo cuvier*. In: The IUCN Red List of Threatened Species 2009, International Union for Conservation of Nature (IUCN), e.T39378A10220026.

Simpfendorfer, C., Burgess, G.H., 2009. *Carcharhinus leucas*. In: The IUCN Red List of Threatened Species 2009, International Union for Conservation of Nature (IUCN), e.T39372A10187195.

Simpfendorfer, C., Compagno, L.J.V., 2015. *Megachasma pelagios*. In: The IUCN Red List of Threatened Species 2015, International Union for Conservation of Nature (IUCN), e.T39338A2900476.

Simpfendorfer, C.A., Milward, N.E., 1993. Utilization of a tropical bay as a nursery area by sharks of the families Carcharhinidae and Sphyrnidae. Environ. Biol. Fishes 37, 337–345.

Skomal, G.B., Hoyos-Padilla, E.M., Kukulya, A., Stokey, R., 2015. Subsurface observations of white shark Carcharodon carcharias predatory behaviour using an autonomous underwater vehicle. J. Fish Biol. 87, 1293–1312.

Smale, M.J., 2009. *Triaenodon obesus*. In: The IUCN Red List of Threatened species 2009, International Union for Conservation of Nature (IUCN), e.T39384A10188990.

Smale, M.J., Cliff, G., 1998. Cephalopods in the diets of four shark species (*Galeocerdo cuvier*, *Sphyrna lewini*, *S. zygaena* and *S. mokarran*) from Kwa Zulu-Natal, South Africa. South Afr. J. Mar. Sci. 20 (1), 241–253.

Smale, M.J., Cliff, G., 2012. White sharks and cephalopod prey. Indicators of habitat use? In: Domeier, M.L. (Ed.), Global Perspectives on the Biology and Life History of the White Shark. Florida. CRC Press, Boca Raton, pp. 51–57.

Smart, J.J., Paul, L.J., Fowler, S.L., 2016. *Chlamydoselachus anguineus*. In: The IUCN Red List of Threatened Species 2016, International Union for Conservation of Nature (IUCN), e. T41794A68617785.

Smith, W.D., Márquez-Farias, J.F., Pérez-Jiménez, J.C., 2009a. *Rhizoprionodon longurio*. In: The IUCN Red List of Threatened Species 2009, International Union for Conservation of Nature (IUCN), e.T161662A5475392.

Smith, W.D., Bizzarro, J.J., Cailliet, G.M., 2009b. The artisanal elasmobranch fishery of Baja California, Mexico: characteristics and management considerations. Cienc. Mar. 35, 209–236.

Soriano-Velásquez, S.R., Acal-Sánchez, D., Castillo-Géniz, J.L., Vázquez-Gómez, N., Ramírez-Santiago, C.E., 2006. Tiburón del Golfo de Tehuantepec. In: Arreguín-Sánchez, F., Beléndez-Moreno, L.F.J., Méndez-Gómez-Humarán, I., Solana-Sansores,-R., Rangel-Dávalos, C. (Eds.), Sustentabilidad y Pesca Responsable en México. Evaluación y Manejo. INAPESCA, México, pp. 323–363.

Sosa-Nishizaki, O., Marquez-Farias, J.F., Villavicencio-Garayzar, C.J., 2008. Case study: pelagic shark fisheries along the west coast of Mexico. In: Camhi, M.D., Pikitch, E.K., Babcock, E.A. (Eds.), Sharks of the Open Ocean: Biology, Fisheries and Conservation. Blackwell Science, Ames, IA. U.S.A, pp. 60–68.

Stevens, J., 2009. *Prionace glauca*. In: The IUCN Red List of Threatened Species 2009, International Union for Conservation of Nature (IUCN), e.T39381A10222811.

Stevens, J.D., Bonfil, R., Dulvy, N.K., Walker, P.A., 2000. The effects of fishing on sharks, rays, and chimaeras (Chondrichthyans), and the implications for marine ecosystems. ICES J. Mar. Sci. 57, 476–494.

Sundström, L.F., 2015. *Negaprion brevirostris*. In: The IUCN Red List of Threatened Species 2015, International Union for Conservation of Nature (IUCN), e.T39380A81769233.

Tamburin, E., Hoyos-Padilla, E.M., Galvan-Magaña, F., Kim, S., Sánchez-González, A., Hernández-Herrera, A., Castillo-Geniz, J.L., Godinez-Padilla, C.J., Elorriaga-Verplancken, F., 2019. Isotopic niche and resource sharing among young white sharks

(*Carcharodon carcharias*) and shortfin mako sharks (*Isurus oxyrinchus*) from Baja California, Mexico. Mar. Ecol. Prog. Ser. 613, 107–124.

Tapia-Garcia, M., Garcia-Abad, M.C., Carranza-Edwards, A., Vazquez-Gutierrez, F., 2007. Environmental Characterization of the Continental Shelf of the Gulf of Tehuantepec, Mexico. Geofisica Internacional (2007). vol. 46. pp. 249–260, Geofisica Internacional, Num. 4.

Taylor, J.G., 1994. Whale Sharks: The Giants of Ningaloo Reef. Harper Collins, New York.

Taylor, J.G., 2007. Ram filter-feeding and nocturnal feeding of whale sharks (*Rhincodon typus*) at Ningaloo reef, Western Australia. Fish. Res. 84, 65–70.

Torres-Huerta, A.M., 2004. Distribución, abundancia y hábitos alimentarious de juveniles del tiburón martillo *Sphyrna lewini* Griffith y Smith (SPHYRNA) en la costa de Sinaloa, México durante el evento El Niño 1997-98. Universidad del Mar, Puerto Angel, Oaxaca, México. M.S. Thesis, 91 pp.

Torres-Rojas, Y.E., Hernandez-Herrera, A., Galvan-Magaña, F., 2006. Feeding habits of the scalloped hammerhead *Sphyrna lewini*, in Mazatlán waters, southern gulf of California, Mexico. Cybium 30, 85–90.

Torres-Rojas, Y.E., Hernandez-Herrera, A., Galvan-Magaña, F., Alatorre-Ramirez, V.G., 2009. Stomach content analysis of juvenile scalloped hammerhead shark *Sphyrna lewini* captured off the coast of Mazatlán, Mexico. Aquat. Ecol. 44, 301–308.

Torres-Rojas, Y.E., Paez-Osuna, F., Hernandez-Herrera, A., Galvan-Magaña, F., Aguiñiga-Garcia, S., Villalobos-Ortiz, H., Sampson, L., 2013. Feeding grounds of juvenile scalloped hammerhead sharks (*Sphyrna lewini*) in the south-eastern Gulf of California. Hydrobiologia 726, 81–94.

Torres-Rojas, Y.E., Páez-Osuna, F., Camalich, J., Galván-Magaña, F., 2015. Diet and trophic level of scalloped hammerhead shark (*Sphyrna lewini*) from the Gulf of California and gulf of Tehuantepec. Mex. Iran. J. Fish. Sci. 14 (3), 767–785.

Tricas, T.C., McCosker, J.E., 1984. Predatory behavior of the white shark (*Carcharodon carcharias*), with notes on its biology. Proc. Calif. Acad. Sci. 43, 221–238.

Valdivia, M.M.A., 2013. Nivel trófico del tiburón angelito *Squatina californica* (Ayres, 1859) mediante el uso de isotopos estables de nitrógeno y carbono en Baja California Sur, México. Bachelor thesis. Universidad del Mar, Oaxaca.

Valenti, S.V., 2009. *Cephalurus cephalus*. In: The IUCN Red List of Threatened Species 2009, International Union for Conservation of Nature (IUCN), e.T161455A5428230.

Velasco, T.P.M., 2005. Hábitos alimenticios e isótopos de 13C y 15N del tiburón mako *Isurus oxyrinchus* (Rafinesque, 1810) en la costa occidental de Baja California Sur. M. S. ThesisCICIMAR-IPN.

Villavicencio-Garayzar, C.J., White, C.F., Lowe, C.G., 2015. *Cephaloscyllium ventriosum*. In: The IUCN Red List of Threatened Species 2015, International Union for Conservation of Nature (IUCN), e.T60227A80671800.

Walker, T.I., Cavanagh, R.D., Stevens, J.D., Carlisle, A.B., Chiaramonte, G.E., Domingo, A., Ebert, D.A., Mancusi, C.M., Massa, A., McCord, M., Morey, G., Paul, L.J., Serena, F., Vooren, C.M., 2006. *Galeorhinus galeus*. In: The IUCN Red List of Threatened Species 2006, International Union for Conservation of Nature (IUCN), e. T39352A10212764.

Weighmann, S., 2016. Annotated checklist of the living sharks, batoids and chimaeras (Chondrichthyes) of the world, with a focus on biogeographical diversity. J. Fish Biol. 88, 837–1037.

Wetherbee, B., Cortés, E., 2004. Food consumption and feeding habits. In: Carrier, J.C., Musick, J.A., Heithaus, M.R. (Eds.), Biology of Sharks and their Relatives. CRC Press, Boca Raton, pp. 225–246.

Wilson, S.G., Taylor, J.G., Pearce, A.F., 2001. The seasonal aggregation of whale sharks at Ningaloo reef, Western Australia: currents, migrations and the El Niño/southern oscillation. Environ. Biol. Fishes 61, 1–11.

Wolfson, F.H., 1987. The Whale Shark, *Rhincodon typus*, Smith 1828, off Baja California, México (Chondrichthyes:Rhinodontidae). In: Memorias del V Simpiosio de Biología Marina. UABCS, La Paz, Mexico, pp. 103–108.
Young, E., 2001. State intervention and abuse of the commons: fisheries development in Baja California Sur, Mexico. Ann. Assoc. Am. Geogr. 91, 283–306.
Zaytsev, O., Cervantes-Duarte, R., Sanchez-Montante, O., Gallegos-Garcia, A., 2003. Coastal upwelling activity on the Pacific shelf of the Baja California peninsula. J. Oceanogr. 59, 489–502.

## Further reading

Baum, J.K., Myers, R.A., 2004. Shifting baselines and the decline of pelagic sharks in the Gulf of Mexico. Ecol. Lett. 7, 135–145.
Becerril-García, E.E., Hoyos-Padilla, E.M., Petatán-Ramírez, D., Galván-Magaña, F., 2019. Southernmost record of the white shark *Carcharodon carcharias* (Chondrichthyes: Lamnidae) in the Mexican Pacific. Lat. Am. J. Aquat. Res. 47 (1), 190–193.
Ferretti, F., Myers, R.A., Serena, F., Lotze, H.K., 2008. Loss of large predatory sharks from the Mediterranean Sea. Conserv. Biol. 22, 952–964.
Furlong-Estrada, E., Tovar-Ávila, J., Ríos-Jara, E., 2014. Evaluación de riesgo ecológico de los tiburones capturados por la pesquería artesanal en la entrada del Golfo de California. Hidrobiológica 24 (2), 83–97.
Furlong-Estrada, E., Tovar-Ávila, J., Pérez-Jiménez, J.C., Ríos-Jara, E., 2015. Resilience of *Sphyrna lewini*, *Carcharhinus falciformis* and *Rhizoprionodon longurio* in the entrance of the Gulf of California after three decades of exploitation. Cienc. Mar. 41 (1), 49–63.
Jorgensen, S.J., Klimley, A.P., Muhlia-Melo, A., 2009. Diving of scalloped in anoxic zone in Gulf of California. J. Fish Biol. 74, 1682–1687.
Kato, S., 1965. White shark, *Carcharodon carcharias*, from the Gulf of California, with a list of the sharks in Mazatlan, Mexico. Copeia 1965, 384.
Klimley, A.P., 1999. Sharks beware. Am. Sci. 87, 488–491.
Klimley, A.P., Ainley, D.G., 1996. White shark research in the past: a prespective. In: Kimley, A.P., Ainley, D.G. (Eds.), Great White Sharks: The Biology of *Carcharodon carcharias*. Academic Press, San Diego, CA, pp. 3–4.
McCosker, J.E., Lea, R.L., 1996. White shark attacks in the eastern Pacific Ocean, an update and analysis. In: Kimley, A.P., Ainley, D.G. (Eds.), Great White Sharks: The Biology of *Carcharodon carcharias*. Academic Press, San Diego, CA, pp. 419–434.
Pérez-Jiménez, J.C., Sosa-Nishizaki, O., Furlong-Estrada, E., Corro-Espinosa, D., Venegas-Herrera, A., Barragán-Cuencas, O.V., 2005. Artisanal shark fishery at "Tres Marias" islands and Isabel Island in the central Mexican Pacific. J. Northwest Atl. Fish. Sci. 35, 333–343.
Santana-Hernández, H., Flores-Valdez, J., 2014. Pelágicos mayores obtenidos por la flota palangrera de mediana altura del puerto de Manzanillo, Colima. INAPESCA, México. 63 p.
Springer, S., 1967. Social organization of shark populations. In: Gilbert, P.W., Matheson, R.F., Rall, D.P. (Eds.), Sharks, skates and Rays. John Hopkins Press, Baltimore, MD, pp. 149–174.
Tamburin, E., Hoyos-Padilla, M., Sanchez-González, A., Hernández-Herrera, A., Elorriaga-Verplancken, F.R., Galván-Magaña, F., 2020. New nursery area for white sharks (*Carcharodon carcharias*) in the eastern Pacific Ocean. Turkish J. Fish. Aquat. Sci. 19 (4).
Torres-Herrera, M.R., Tovar-Ávila, J., 2014. Variación temporal de la captura de tiburón en las islas y costa central de Nayarit, México, con base en los registros oficiales de desembarque. Hidrobiológica 24 (2), 99–107.

Tovar-Ávila, J., Patiño-Valencia, J.L., Zárate-Becerra, M.E., 2017. Análisis de la captura de tiburón en la Isla Isabel, Nayarit, México, durante el periodo 2007-2013. Cienc. Pesq. 25 (1), 5–14.

Tovar-Ávila, J., Rodríguez-Madrigal, J.A., Mondragón-Sánchez, L.F., Aviña-Romero, M.J., Torres-Herrera, M.R., Chávez-Arrenquín, D.A., Álvarez-Fuentes, C.J., Santana-Hernández, H., Jacob-Cervantes, M., Hernández-Ventura, S., Espinosa-García, A., Marín-Zaldívar, F.L., Zárate-Becerra, M.E., Acosta-Quintana, J.D., García-Meraz, A., Patiño-Valencia, J.L., 2018. Pesquería y dinámica poblacional de los elasmobranquios en la entrada al Golfo de California y el Pacífico Central Mexicano. Informe de investigación, INAPESCA, México. 59 p.

CHAPTER FOUR

# Conservation genetics of elasmobranchs of the Mexican Pacific Coast, trends and perspectives

**Jonathan Sandoval-Castillo***
Molecular Ecology Lab, College of Science and Engineering, Flinders University, Adelaide, SA, Australia
*Corresponding author: e-mail address: jonathan.sandoval-castillo@flinders.edu.au

## Contents



## Abstract

One of the most critical threats to biodiversity is the high extinction rate driven by human activities. Reducing extinction rates requires the implementation of conservation programmes based on robust scientific data. Elasmobranchs are important ecological components of the ocean, and several species sustain substantial economic activities. Unfortunately, elasmobranchs are one of the most threatened and understudied animal taxa. The Mexican Pacific Coast (MPC) is a region with high elasmobranch diversity and is the seat of major elasmobranch fisheries. But it is also a developing region with several conservation and management challenges which require national and international attention. Here, we review the conservation genetics

*Advances in Marine Biology*, Volume 83
ISSN 0065-2881
https://doi.org/10.1016/bs.amb.2019.08.002

literature of elasmobranchs from the MPC. We present a synthesis of the works using samples from the region and emphasize the main gaps and biases in these data. In addition, we discuss the benefits and challenges of generating genomic information to improve the management and conservation of an elasmobranch biodiversity hotspot in a developing country. We found 47 elasmobranch genetic articles that cover <30% of the elasmobranch diversity in the region. These studies mainly used mitochondrial DNA sequences to analyse the genetic structure of commercially important and abundant species of the order Carcharhiniformes. Some of these papers also assessed mating systems, demographic parameters, and taxonomic uncertainties, all of which are important topics for efficient management decisions. In terms of conservation genetics, elasmobranchs from the MPC remain understudied. However, high-throughput sequencing technologies have increased the power and accessibility of genomic tools, even in developing countries such as Mexico. The tools described here provide information relevant for biodiversity conservation. Therefore, we strongly suggest that investment in genomic research will assist implementation of efficient management strategies. In time, this will reduce the extinction risk of the unique elasmobranch biodiversity from the MPC.

## 1. Background

Biological extinctions are natural phenomena, and five mass extinctions have occurred before human existence. However, the current rate of diversity loss is higher than pre-human rates (Ceballos et al., 2017). This is considered one of the most critical threats to ecosystem health and services and consequently to human well-being (Ceballos et al., 2017; Young et al., 2016). Several species and innumerable populations of animals are already extinct as direct or indirect consequence of human actions, while many more are being driven to extinction (Ceballos et al., 2015; Young et al., 2016). Reducing anthropogenic defaunation rates requires the implementation of conservation and management programmes based on robust scientific data about the taxonomy, ecology and conservation status of the global fauna. However, our understanding of the basic biology of most animal species is inadequate. Moreover, the existing information is both taxonomically (Donaldson et al., 2016) and geographically biased (Beheregaray, 2008; Meijaard et al., 2015). These biases are particularly true for sharks and rays. Despite the fact that several elasmobranchs are large, charismatic and economically important, the taxonomic group is underrepresented in the scientific literature (Domingues et al., 2018; Dulvy et al., 2014; Heupel and Simpfendorfer, 2010; McClenachan et al., 2012; IUCN). This lack of information about elasmobranchs is more evident in developing nations

(Bester-van der Merwe and Gledhill, 2015; Dulvy et al., 2014; Reis et al., 2016; White and Kyne, 2010), where elasmobranch diversity is relatively high (see chapter "Biodiversity and Conservation of sharks in Pacific Mexico" by Saldaña Ruiz et al. this volume; Stein et al., 2018) and is coupled with greater fishery pressures and management challenges (Davidson et al., 2016; Dulvy et al., 2017; Kyne et al., 2012; Worm et al., 2013; Worm and Branch, 2012).

Despite this lack of information, elasmobranchs have been and remain of particular interest to conservation researchers. They have a unique evolutionary history, with long-term high diversity due to several radiations spanning the last ~400 million years (Compagno, 1990; Stein et al., 2018). The group has sophisticated morphological, ecological and behavioural adaptive specializations, which have allowed colonization of several freshwater and almost all marine habitats (Compagno, 1990). As predators, elasmobranchs have important roles as connectors throughout food webs, across habitats and ecosystems (see chapter "Shark ecology, the role of the apex predator and current conservation status" by Galván-Magaña et al. this volume; Wetherbee and Cortés, 2004), where they export and import energy and nutrients between different level of the trophic webs, or between different ecosystems. They can also be key in shaping community and ecosystem structures, and changes in their abundance could have strong ecological effects, especially in the marine realm (Bond et al., 2019; Bornatowski et al., 2018; Grubbs et al., 2016; Stevens et al., 2000). Historically, elasmobranchs have been an important income and food resource for humans (Applegate et al., 1993; Rick et al., 2002), and in the last decades we have witnessed the major expansion of shark and ray fisheries at a global scale (Dent and Clarke, 2015; see chapter "Fisheries interactions and the challenges of both targeted and non-targeted take in shark conservation" by Sosa-Nishizaki of volume 84). More recently, wild elasmobranch sighting tourism has also increased around the word. This activity has the potential to generate increased economic, conservation and social benefits (Mieras et al., 2017). However, this industry is underdeveloped and focuses on few species, leaving fisheries as the main economic importance of elasmobranchs (see chapter "The economy of shark conservation: The role of ecotourism and citizen science" by Cisneros-Montemayor of volume 84; Plata Zepeda et al., 2018). Unfortunately, many elasmobranchs show extreme life histories, including late maturity, long gestation periods and low fecundity, making them intrinsically susceptible to overexploitation (Dulvy et al., 2017). Hence, numerous elasmobranch populations have diminished due to overfishing (see chapter "Biodiversity and Conservation of sharks in Pacific Mexico" by Saldaña Ruiz et al. this volume; Camhi et al., 2009; Dulvy et al., 2017; Oliver et al., 2015; see chapter "Fisheries interactions and the challenges of both targeted and non-targeted take in shark conservation" by Sosa-Nishizaki of volume 84). In fact, around 31% of elasmobranch species are threatened with the risk of extinction, making them globally one of the most threatened animal taxa (Dulvy et al., 2014; IUCN).

The intrinsically limited accessibility of marine habitats makes problematic the collection of direct biological observations for marine animals, an issue which is emphasized in highly mobile and low-density species such as elasmobranchs. Molecular approaches provide a unique tool to solve these problems. With relatively few small tissue samples collected from live or dead animals, we can extract genomic data and obtain biological information that might be unreachable by other means (Frankham et al., 2010). From individual relationships (parentage or relatedness) and population demography (migration, population size, demographic history, population structure and mating systems), to phylogenetic relationships (taxonomic status) and evolutionary processes (adaptation, introgression, responses to climatic change), the potential information generated by genomic methods is fundamental to the development of efficient conservation plans (Allendorf et al., 2010; Frankham et al., 2010; Primmer, 2009). The use of genetic data to assist conservation and management has been so important in recent decades that conservation genetics has emerged as a new research discipline (Frankham et al., 2010). This discipline integrates evolutionary genetic theory and molecular tool technologies with the main goal of preventing population extirpations and consequent species extinctions, therefore preserving the ecological and evolutionary processes in which these biological entities are involved (Frankham et al., 2010; Hedrick and Miller, 1992). The increasing use of and advances in conservation genetic approaches have been reflected by better conservation management of some elasmobranch species in the last decade (Dudgeon et al., 2012; Larson et al., 2017). Despite the potential benefits of genetic information on shark and ray conservation, genetic data is absent for most elasmobranch species. Furthermore, most conservation efforts currently do not consider existing studies (Domingues et al., 2018; Ovenden et al., 2015). For example, <20% of elasmobranch species in Mexico haven been investigated in terms of genetic/genomics analyses, and the current elasmobranch fisheries management law (NOM-29-PESC-2006; DOF 2018) does not include any reference to these investigations.

To date, some reviews have described the approaches and findings of elasmobranch conservation genetics (Domingues et al., 2018; Dudgeon et al., 2012; Larson et al., 2017), however, none of them have focussed on Mexican elasmobranch fauna. Mexico is recognized as an elasmobranch biodiversity hotspot, a major source of elasmobranch products and a priority region for the conservation of sharks and rays (Dulvy et al., 2017; Stein et al., 2018). This is particularly true for the Mexican Pacific Coast (MPC), where ~120 species of sharks and rays have been identified (Ehemann et al., 2018) and over 5000 tonnes of elasmobranch catches are reported every year (see chapters "Biodiversity and Conservation of sharks in Pacific Mexico" by Saldaña Ruiz et al. this volume; "Fisheries interactions and the challenges of both targeted and non-targeted take in shark conservation" by Sosa-Nishizaki of volume 84). Here, we provide a long-needed review of elasmobranch genetic conservation from the MPC based on literature available previous to February 2019. This review has three broad objectives: First, we attempt to present a numerical synthesis of the work done with samples from the MPC, including species, markers, geographic area, and main research questions. Second, we emphasize the strong taxonomic, geographic and institutional bias in conservation genetic research, which, although not exclusive to the MPC, is highly evident in the region. Third, we highlight the enormous gap in elasmobranch conservation genetics of the MPC and discuss the benefits and challenges of generating genomic information to assist with the management and conservation of an elasmobranch biodiversity hotspot in a developing country. We expect the review will promote collaborative research efforts intended to reduce bias and fill gaps in elasmobranch conservation genetic research from the region.

## 2. Studies to date

For this review we compiled a database by searching in Web of Science and Google Scholar using the terms 'genetics', 'genomics', 'elasmobranch', 'shark', 'ray', 'Pacific', and 'Mexico' in the title, abstract and keywords section of all articles published before February 2019. From there, we scanned all results and selected papers where samples from MPC were used. This resulted in a list of 47 scientific papers on which we base our review.

### 2.1 Markers

The use of molecular methods in elasmobranchs from the MPC has followed the development of the field at the global scale. This chapter is by no means an exhaustive review of conservation genetics methodologies used in elasmobranchs, since other recently published reviews have described them extensively (Dudgeon et al., 2012; Larson et al., 2017).

Rather, we discuss these approaches as they have been specifically applied in the MPC, as well as their context within the broader research.

Allozyme markers dominated genetic studies of sharks around the world during the 1980s and 90s (Dudgeon et al., 2012). However, this was not the case in the MPC. Only a single study, and the first published example of shark genetic research in the MPC, used allozymes to distinguish between species of thresher sharks (Eitner, 1995). The later development of polymerase chain reaction (PCR) technology facilitated the use of specific regions from both the nuclear and the mitochondrial genomes in genetic research. Due to the ease of isolation and its relatively low mutational rate, mitochondrial DNA (mtDNA) markers were extensively used in MPC elasmobranch research. Particularly in phylogeographic and population structure analyses either with sequences (Félix-López et al., 2019; Kitamura et al., 1996) or restriction fragment length polymorphisms (RFLP's; Grijalva-Chon et al., 2002). Meanwhile, the use of nuclear markers has allowed for higher resolution analyses and the exploration of more recent evolutionary and ecological forces. These range from contemporary connectivity between populations to individual relatedness. Research using amplification of nuclear fragments included one sequence-based (Castillo-Páez et al., 2017), two RFLP-based (Pérez Jiménez et al., 2005; Pérez-Jiménez et al., 2013) and several microsatellite-based (Bernard et al., 2018; Byrne and Avise, 2012; Schultz et al., 2008; see Table 1) studies. Additionally, an unpublished work used amplified fragment length polymorphism (AFLPs; Sandoval-Castillo and Beheregaray, 2019). It should be noted that a combination of both nuclear and mitochondrial markers can allow the integration of contemporary and historical perspectives, providing greater power for interpretation of evolutionary and ecological patterns. However, <30% of the listed studies incorporate both types of markers (Table 1).

The most recent advances in sequencing technologies are now allowing massive sequence data production at a very affordable cost, even for non-model species (Luikart et al., 2019). This has shifted genetic analysis in two main ways. First, there has been a shift from the use of only a few markers per study, to the use of thousands of markers distributed genome-wide, to the use of entire genomes (Attard et al., 2018) which have changed radically the possibilities of genomic analyses. Second, there has been a shift from the use of only putatively neutral markers to the exploration of adaptive genetic variation (Sandoval-Castillo et al., 2018). This transition has been slow within the study of elasmobranchs but has been demonstrated in at least three studies from the MPC, including the use of the whole mitochondrial

**Table 1** Genetic studies in elasmobranchs from the Mexican Pacific Coast.

| Family | Species | Genetic markers | Locality within the Mexican Pacific Coast | Geographic extension | mtDNA diversity | nDNA diversity | Citation |
|---|---|---|---|---|---|---|---|
| Alopidea | *Alopias pelagicus* | mtDNA(COI) | Mexican TNP | Pacific Ocean | h = 0.59 $\pi$ = 0.10 | | Cardeñosa et al. (2014) |
| | *Alopias* spp. | 13 Allozymes | Mexican TNP | Northeast Pacific | | | Eitner (1995) |
| | *Alopias* spp. | mtDNA(CR) | Mexican TNP | Global | | | Trejo (2005) |
| Carcharhinidae | *Carcharhinus brachyurus* | mtDNA(CR) | Baja California Peninsula | Global | h = 0.76 $\pi$ = 1.60 | | Benavides et al. (2011) |
| | *Carcharhinus falciformis* | mtDNA(CR) | Mexican Pacific Coast | Pacific Ocean | h = 0.48 $\pi$ = 0.09 | | Galván-Tirado et al. (2013) |
| | *Carcharhinus falciformis* | mtDNA(CR) | Mexican Pacific Coast | Global | h = 0.93 $\pi$ = 0.61 | | Clarke et al. (2015) |
| | *Carcharhinus galapagensis* | SNPs | Revillagigedo Islands | Interoceanic | h = 0.69 $\pi$ = 3.08 | He = 0.21 | Pazmiño et al. (2018) |
| | *Carcharhinus galapagensis* *Carcharhinus obscurus* | mtDNA(ND2) 900 exons | Gulf of California Revillagigedo Islands | Mexican Pacific Coast | h = 0.89 $\pi$ = 0.5 | | Corrigan et al. (2017) |
| | *Carcharhinus leucas* | mtDNA(CR/Ctb) | Mexican TEP | Interoceanic | | | Kitamura et al. (1996) |
| | *Carcharhinus limbatus* | mtDNA(CR) | Gulf of California | Global | h = 0.84 $\pi$ = 0.41 | | Keeney and Heist (2006) |
| | *Carcharodon carcharias* | mtDNA(CR) | Guadalupe Island | Pacific Ocean | h = 0.79 $\pi$ = 0.31 | | Jorgensen et al. (2009) |
| | *Negaprion brevirostris* | mtDNA(CR)/9 Mst | Mexican TNP | Global | h = 0.78 $\pi$ = 0.58 | Ho = 0.73 He = 0.81 | Schultz et al. (2008) |
| | *Prionace glauca* | 14 Mst | Baja California Peninsula | Pacific Ocean | | Ho = 0.62 He = 0.60 | King et al. (2015) |

*Continued*

**Table 1** Genetic studies in elasmobranchs from the Mexican Pacific Coast.—cont'd

| Family | Species | Genetic markers | Locality within the Mexican Pacific Coast | Geographic extension | mtDNA diversity | nDNA diversity | Citation |
|---|---|---|---|---|---|---|---|
| Gymnuridae | *Gymnura marmorata* *Gymnura crebripunctata* | mtDNA (Ctb) | Mexican TNP | Mexican TNP | | | Smith et al. (2009) |
| Hexanchidae | *Hexanchus griseus* | mtDNA(CR/Ctb/16S) | Mexican TNP | Global | $h=0.83$ $\pi=0.10$ | | Vella and Vella (2017) |
| Lamnidae | *Carcharodon carcharias* | mtDNA(CR) | Guadalupe Island | Global | $h=1$ $\pi=0.04$ | | Gubili et al. (2012) |
| | *Carcharodon carcharias* | MtDAN(CR) | Mexican TNP | Interoceanic | $h=0.77$ $\pi=0.18$ | | Oñate-González et al. (2015) |
| | *Carcharodon carcharias* | mtDNA(Genome) | Baja California Peninsula | Baja California Peninsula | $h=0.73$ $\pi=0.13$ | | Díaz-Jaimes et al. (2016) |
| | *Carcharodon carcharias* | 10 Mst | Guadalupe Island | Pacific Ocean | | $He=0.58$ | Bernard et al. (2018) |
| Megachasmidae | *Megachasma pelagios* | mtDNA(Cox1)/1 Mst | Baja California Peninsula | Pacific Ocean | $h=0.33$ $\pi=0.06$ | | Liu et al. (2018) |
| Myliobatidae | *Rhinoptera steindachneri* | mtDNA(ND2) | Mexican TNP | Mexican TNP | h-0.54 $\pi=5.30$ | | Sandoval-Castillo and Rocha-Olivares (2011) |
| | *Mobulidae* | mtDNA(Whole Genome) 1000 exons | Mexican TNP | Global | | | White et al. (2017) |
| Pristidae | *Pristis pectinata Pristis pristis* | eDNA | Mexican TEP | Mexican TEP | | | Bonfil (Personal communication) |

| | | | | | | | |
|---|---|---|---|---|---|---|---|
| Rhincodontidae | *Rhincodon typus* | mtDNA(CR) | Gulf of California | Interoceanic | h = 0.96 $\pi$ = 0.61 | | Castro et al. (2007) |
| | *Rhincodon typus* | mtDNA(CR) | Gulf of California | Gulf of California | h = 0.90 $\pi$ = 0.50 | | Ramírez-Macías et al. (2007) |
| | *Rhincodon typus* | 8 Mst | Gulf of California | Global | | Ho = 0.66 He = 0.69 | Schmidt et al. (2009) |
| | *Rhincodon typus* | mtDNA(CR)/14 Mst | Gulf of California | Global | h = 0.92 $\pi$ = 1.2 | Ho = 0.64 He = 0.60 | Vignaud et al. (2014) |
| Rhinobatidae | *Pseudobatus productus* | RFLP | Mexican TNP | Mexican TNP | h = 0.77 $\pi$ = 1.19 | | Sandoval-Castillo et al. (2004) |
| | *Zapteryx exasperata* | mtDNA(ND2/CR/CAT) | Mexican TNP | Mexican TNP | h = 0.76, 0.39,0.84 $\pi$ = 0.13, 0.07, 0.11 | | Castillo-Páez et al. (2013) |
| | *Zapteryx exasperata* *Zapteryx xyster* | mtDNA(ND2, CR, CR) nDNA (RAG1) | Mexican Pacific Coast | Mexican Pacific Coast | | | Castillo-Páez et al. (2017) |
| | *Pseudobatus productus* | mtDNA(CR/ND2) 500 AFLP | Mexican TNP | Mexican TNP | | | Sandoval-Castillo and Beheregaray (2019) |
| Sphyrnidae | *Sphyrna lewini* | mtDNA(CR) | Mexican TNP | Global | h = −0.51 $\pi$ = 0.09 | | Duncan et al. (2006) |
| | *Sphyrna lewini* | mtDNA(CR)/15 Mst | Mexican TNP | Northeast Pacific | h = 0.53 $\pi$ = 0.11 | Ho = 0.77 He = 0.79 | Nance et al. (2011) |
| | *Sphyrna lewini* | mtDNA(CR)/5 Mst | Mexican Pacific Coast | Mexican Pacific Coast | h = 0.49 $\pi$ = 1.10 | | Castillo-Olguín et al. (2012) |
| | *Sphyrna lewini* | 6 Mst | Baja California Peninsula | Global | | Ho = 0.70 He = 0.75 | Daly-Engel et al. (2012) |
| | *Sphyrna zygaena* | mtDNA (CR) | Mexican TNP | Mexican TNP | | | Félix-López et al. (2019) |

*Continued*

**Table 1** Genetic studies in elasmobranchs from the Mexican Pacific Coast.—cont'd

| Family | Species | Genetic markers | Locality within the Mexican Pacific Coast | Geographic extension | mtDNA diversity | nDNA diversity | Citation |
|---|---|---|---|---|---|---|---|
| Squatinidae | *Squatina californica* | RFLP | Gulf of California | Gulf of California | h = 0.14 π = 0.08 | | Grijalva-Chon et al. (2002) |
| | *Squatina* spp. | mtDNA(COI/16S) | Mexican TNP | Global | | | Stelbrink et al. (2010) |
| | *Squatina californica* | mtDNA(CR) | Mexican TNP | Mexican TNP | h = 0.97 π = 1.3 | | Ramírez-Amaro et al. (2017) |
| Triakidae | *Mustelus albipinnis* | RFLP/ITS1 | Mexican TNP | Mexican TNP | | | Pérez Jiménez et al. (2005) |
| | *Mustelus californicus* | 4 Mst | Gulf of California | Gulf of California | | | Tarula-Marin and Saavedra-Sotelo (2019) |
| | *Mustelus henlei* | 4 Mst | Baja California Peninsula | Baja California Peninsula | | | Byrne and Avise (2012) |
| | *Mustelus henlei* | 4 Mst | Mexican Pacific Coast | Northeast Pacific | | | Chabot and Haggin (2014) |
| | *Mustelus henlei* | mtDNA(CR)/6 Mst | Mexican TNP | Northeast Pacific | h = 0.77 π = 0.40 | Ho = 0.45 He = 0.56 | Chabot et al. (2015) |
| | *Mustelus henlei* | mtDNA(CR)/6 Mst | Mexican TNP | Mexican TNP | h = 0.84 π = 0.33 | Ho = 0.71 He = 0.68 | Sandoval-Castillo and Beheregaray (2015) |
| | *Mustelus spp* | RFLP/ITS2 | Mexican TNP | Mexican TNP | | | Pérez-Jiménez et al. (2013) |
| | *Ttriakis semifasciata* | mtDNA (CR)/5 Mst | Baja California Peninsula | Northeast Pacific | | Ho = 0.80 He = 0.81 | Barker et al. (2015) |
| Urolophidae | *Urobatis halleri* | 7Mst | Mexican TNP | Northeast Pacific | | He = 0.88 | Plank et al. (2010) |

AFLP = amplified fragment length polymorphisms; CO1 = cytochrome oxidase 1; CR = control region; Ctb = cytochrome *b*; eDNA = environmental DNA; ITS1 = Internal transcribed spacer 1; Mst = microsatellites; mtDNA = mitochondrial DNA; NCR = non-coding region; nDNA = nuclear DNA; ND2 = NADH-dehydrogenase 2; RAG1 = recombination activating gene 1; RFLP = restriction fragment length polymorphisms; SNP = single nucleotide polymorphisms; WG = whole genome; 16S = 16S ribosomal RNA; TNP = Temperate Northern Pacific; TEP = Tropical Eastern Pacific; h = haplotype diversity; π = nucleotide diversity; Ho = observed heterozygosity; He = expected heterozygosity.

genome (Corrigan et al., 2017), hundreds of exons (White et al., 2017) and thousands of single nucleotide polymorphisms (SNPs) (Pazmiño et al., 2018). Not only do these changes improve resolution for delimitations of taxonomic units, but they also allow the development of management plans that consider adaptive potential under scenarios of future climatic change.

## 2.2 Applications

As elasmobranch conservation genetics is a relatively young area of research, we briefly review how molecular methods have been applied to studies of sharks and rays from the MPC. We discuss applications ranging from the broad taxonomic to the intraspecific level, and their relevance for elasmobranch conservation.

### *2.2.1 Taxonomic delimitation and species identification*

Accurate species identification is fundamental for the implementation of efficient management and conservation plans. Often, morphology is the easiest approach for species identification. However, there are many circumstances where genetic techniques are more efficient, including the identification of morphologically cryptic species, as well as morphologically different life stages, body parts or hybrids.

Morphologically similar or cryptic elasmobranch lineages have been identified in the MPC using genetic approaches. The single study in this region to use allozymes suggested the existence of a cryptic lineage (lineages that are morphologically impossible or extremely difficult to delimit) of the thresher shark Alopias (Eitner, 1995). While thresher shark samples collected from the Pacific Coast of the Baja California Peninsula were morphologically similar to *A. superciliosus*, they were found to be genetically more similar to *A. vulpinus*. Similarly, cryptic allopatric lineages separated by the Baja California Peninsula were described first in the shovelnose guitarfish *Pseudobatus productus* (Sandoval-Castillo et al., 2004) and then in the Pacific cownose ray *Rhinoptera steindachneri* (Sandoval-Castillo and Rocha-Olivares, 2011). In both cases, high genetic distance ($F_{ST} > 0.5$) between two mitochondrial lineages suggested the existence of cryptic species. Other cryptic lineages have been found in samples of the banded guitarfish from the North and South MPC (Castillo-Páez et al., 2017). In this group, the morphological plasticity of the northern lineages partially overlaps with that of the southern lineages and suggests that morphologically diagnostic characteristics of already described lineages are uninformative.

Later works suggest the existence of disparate evolutionary units (populations that are historically isolated and likely to have distinct evolutionary potential) in the pelagic thresher shark *A. pelagicus* from the North Pacific (Cardeñosa et al., 2014), in the great white shark *Carcharodon carcharias* from the Pacific Ocean (Oñate-González et al., 2015) and in the shovelnose guitarfish from the Gulf of California (Sandoval-Castillo and Beheregaray, 2019). In these examples, additional phenotypic and genetic studies are needed to determine the taxonomic status of the cryptic lineages. However, all examples clearly show cryptic evolutionary significant units that would have likely remained unnoticed without the use of genetic approaches. To date, the only compelling genetic evidence that resulted in the description of a new species in the MPC is the work by Pérez Jiménez et al. (2005) and Pérez-Jiménez et al. (2013) They combined RFLPs of nuclear genes with morphological and reproductive data to describe a new species of smooth-hound shark, *Mustelus albipinnis*. While such outcomes have been so far limited, all studies revealing cryptic lineages are important for identifying underestimated biodiversity, and for attributing the correct life history, ecological and demographic traits for improved management of either cryptic species or populations.

Delineation of taxonomic groups is impractical if they cannot be empirically identified. Globally, there are several examples of the use of genetic barcodes on shark body parts and processed fisheries products to identify individuals to the level of species, population or geographic region of origin (Hanner et al., 2016; Rodrigues-Filho et al., 2012; Ward et al., 2008). This approach has proved to be an excellent tool in the management and conservation of elasmobranchs, in two main ways. First, in the international trade of elasmobranch products, visual identification of species is practically impossible. Whole bodies of organisms are not typically available, and, when they are, the necessary taxonomic skills of border protection authorities may be lacking. The use of genetic barcodes has allowed the otherwise impossible identification of international trades of elasmobranch species controlled under the Convention of International Trade in Endangered Species. These include species found in the MPC such as hammerheads sharks *Sphyrna* spp. (Abercrombie et al., 2005), the basking shark *Cetorhinus maximus* (Magnussen et al., 2007) and manta rays *Mobula* spp. (Steinke et al., 2017).

From a more local perspective, correct species identification of catch and by-catch is essential for defining the species composition of fisheries and the delineation and assessment of fishery stocks. However, one of the biggest problems in the management and monitoring of elasmobranch fisheries is the lack of species-specific data on catch and landings (Dulvy et al., 2014; see chapter "Fisheries interactions and the challenges of both targeted and non-targeted take in shark conservation" by Sosa-Nishizaki of volume 84). This problem is highlighted in developing countries like Mexico, where fisheries and management officers, who directly implement the management policies, have limited legal, economic and educational resources necessary to enforce the identification of target or by-catch species on board. For example, in several parts of the MPC, most species of the genus *Carcharhinus* are reported as one species (Sandoval-Castillo personal communication). Worse, juveniles of this genus are reported as 'cazon' together with species from a different family with very different life history and demographic characteristics (Triakidae) and thus with different conservation management priorities. Affordable and efficient genetic assays for elasmobranch species identification have been developed using species-specific primers on a multiplex Polymerase Chain Reaction (PCR), some of which can differentiate and assign samples to up to 21 species (Caballero et al., 2012; Hanner et al., 2016; Rodrigues-Filho et al., 2012). These are rapid, low-cost techniques that require minimum technology, and as such they are suitable for use in developing countries where both economic and technological resources are limited. Unfortunately, none of these techniques have been applied in the MPC, but recent efforts to implement them have been conducted in the Baja California Peninsula (Flores-Ramirez personal communication, 2019). Their routine application in the monitoring of elasmobranch fisheries would provide baseline information necessary for improved management and conservation decisions in the region.

### 2.2.2 Geographic patterns of genetic structure

Exploring the spatial patterns of genetic differentiation within lineages is important for identifying appropriate management units within a geographic jurisdiction (Ovenden et al., 2015). It is also imperative to identify and understand the historical and contemporary processes which create and maintain global biodiversity (Beheregaray, 2008; Ovenden et al., 2015; Rodríguez-Correa et al., 2017). This is vital for effective management of exploited species. Due to the relatively high mobility of most species, it may be anticipated that elasmobranchs would show low genetic differentiation within species' geographic distributions. However, several studies have shown that elasmobranchs vary widely in levels of genetic subdivision, depending on taxonomic group and spatial scale (Dudgeon et al., 2012; Larson et al., 2017). These range from low genetic differentiation between ocean basins in large pelagic species (e.g. Bernard et al., 2018; Jorgensen et al., 2009; King et al., 2015), to strong genetic subdivision at narrow regional scales in small coastal species (e.g. Castillo-Páez et al., 2017; Lewallen et al., 2007; Sandoval-Castillo and Beheregaray, 2015; Sandoval-Castillo et al., 2004). Another general pattern in population studies of elasmobranchs is that nuclear markers tend to show greater genetic connectivity, while mitochondrial markers tend to show stronger genetic differentiation due in part to differential evolutionary mutation rates, inheritance and selection for variability. Since mtDNA is matrilineally inherited while nuclear DNA (nDNA) is both maternally and paternally inherited, the frequently greater genetic divergences in mtDNA is suggestive of a general sex-biased dispersal in elasmobranches. This has been associated with stronger female philopatry and a higher male migratory potential (Dudgeon et al., 2012; Feldheim et al., 2014; Portnoy and Heist, 2012). However, these patterns are not inexorable; even codistributed and closely related species can have very different dispersal patterns (Corrigan et al., 2015a; Phillips et al., 2017).

At present, population structure studies represent the vast majority of the genetic literature from the MPC (see Table 1). Many of these works are at global, interoceanic or oceanic scales, where a few samples from one or two localities on the MPC are included, of relatively large pelagic species. But there are also several papers of a regional scale (Northeast Pacific; NEP), national scale (MPC) or subregional scale (Mexican Northern Temperate Pacific; MNTP; Mexican Tropical Easter Pacific; MTEP), focussed on the genetic populations of mainly coastal epibenthic species and some coastal pelagic species.

#### 2.2.2.1 Global and interoceanic:

At the largest geographic scale, sharks from the Carcharhinidae family dominate research in the MPC, perhaps because many of these species are commercially important and relatively abundant worldwide. Few papers show interoceanic or even global connectivity when using nuclear markers. For example, using eight microsatellites, Schmidt et al. (2009) reported a single population of the whale shark *Rhincodon typus* at global scale. However, when using the mtDNA control region, Vignaud et al. (2014) found a significant differentiation between the Atlantic and Indo-Pacific populations. Similarly, the scalloped hammerhead *Sphyrna lewini* shows mtDNA genetic structure, while nuclear markers show connectivity across the Indian and Pacific Oceans (Daly-Engel et al., 2012; Duncan et al., 2006). These results highlight the importance of incorporating both nuclear and mitochondrial information, to disentangle not just sex-biased, but also demographic and time scale dependent ecological processes which are relevant to the genetic structure of elasmobranchs.

#### 2.2.2.2 The Pacific Ocean:

Three other species show extensive connectivity within the Pacific Ocean, even when using mtDNA; the silky shark *Carcharhinus falciformis* (Clarke et al., 2015; Galván-Tirado et al., 2013), blue shark *Prionace glauca* (King et al., 2015) and megamouth shark *Megachasma pelagios* (Liu et al., 2018). Moreover, the blue shark and the megamouth shark show no apparent division between Atlantic and Pacific Oceans with either mitochondrial or nuclear data (Liu et al., 2018; Veríssimo et al., 2017). However, a lack of genetic structure can be found in two very different demographic scenarios: one with relatively small populations with very high demographic connectivity, and the other with very large but demographically independent populations (Gagnaire et al., 2015). The blue shark is perhaps the most abundant pelagic shark (Compagno et al., 2005), and modelling data have suggested that the lack of genetic differentiation in this species is due to large population size rather than high demographic connectivity (Bailleul et al., 2018). Conversely, the megamouth shark is rarely observed (Compagno et al., 2005), and the low genetic diversity of the species is suggestive of a small population size (Liu et al., 2018). The observed lack of genetic differentiation across ocean basis is therefore most likely due to high connectivity in the megamouth shark. Distinguishing between these two possible scenarios is important for determining patterns of evolutionary resilience due to local adaptation (Allendorf et al., 2010) that should be considered in any long-term management plan.

Most of the population studies show genetic breaks at oceanic level, especially between the Western and Eastern Pacific. The open Pacific Ocean appears to be an effective biogeographic barrier, even for relatively large and highly mobile species such as the great white shark *Carcharodon carcharias* (Bernard et al., 2018; Jorgensen et al., 2009; Oñate-González et al., 2015), the Galapagos shark *Carcharhinus galapagensis* (Pazmiño et al., 2018), and the pelagic thresher shark *Alopias pelagicus* (Cardeñosa et al., 2014). These species show moderate but significant genetic differentiation between Mexico and the Southwest Pacific, a pattern consistent among studies using mtDNA (Cardeñosa et al., 2014; Oñate-González et al., 2015; Vella and Vella, 2017), microsatellites (Bernard et al., 2018; Cardeñosa et al., 2014) and SNPs (Pazmiño et al., 2018). Ambient temperature directly affects the physiological processes of any organism, including elasmobranchs (Gervais et al., 2018; Pouca et al., 2019). At the same time, the thermal environment influences the biological interactions experienced by these organisms, from parasitism (Dallarés et al., 2017) to competition and predation (Yates et al., 2015). Indeed, temperature gradients have been considered important biogeographic barriers, limiting not just the distribution of species (Aguilar et al., 2019; Goodman et al., 2019), but also the gene flow between populations (Sandoval-Castillo et al., 2018). Warm equatorial waters are also considered an important biogeographic barrier for several marine species, including some sharks (Chabot and Allen, 2009). Genetic differentiation between the MPC and South American Pacific coast has so far been suggested only in the copper shark *Carcharhinus brachyurus* (Benavides et al., 2011). However, there is not more comparative analysis with samples from both regions, highlighting the need for more collaborative work between North and South American researchers.

#### 2.2.2.3 Mexican Pacific region:

When reviewing the works at a more regional scale, most species studied at this geographic level are benthic or epibenthic species. The most frequent pattern is the 'allopatric' isolation between the Gulf of California (GC) and the Pacific Coast of the Baja California Peninsula (PBC), which have been reported in the shovelnose guitarfish *Pseudobatos productus* (Sandoval-Castillo et al., 2004); the California butterfly ray *Gymnura marmorata* (Smith et al., 2009); the cownose ray *Rhinoptera steindachneri* (Sandoval-Castillo and Rocha-Olivares, 2011); the banded guitarfish *Zapterix exasperate* (Castillo-Páez et al., 2013); and the California angel shark *Squatina californica* (Ramírez-Amaro et al., 2017). The congruent pattern among these species highlight the importance of the Baja California Peninsula as a biogeographic barrier and the need to consider different evolutionary units for the management of these species.

Two species of sharks form the family Triakidae show genetic separation between North Santa Barbara (California) and the PBC and the GC. These are the leopard shark *Triakis semifasciata* (Barker et al., 2015) and the brown smooth-hound shark *Mustelus henlei* (Chabot et al., 2015). Barker et al. (2015) also report a small differentiation between Mexico and Southern California, but the number of samples used was very small, which can bias the level of genetic differentiation estimated. These sharks have relatively higher mobility than benthic rays from the family Urotrygonidae, however the round stingray *Urobatis halleri* shows genetic homogeneity from the GC to the Southern California coast (Plank et al., 2010). More interestingly, both the round stingray and the brown smooth-hound shark have high connectivity over 2500 km, but also both species show genetic differentiation between Santa Catalina Island and the California Coast (~40 km distance). Both studies suggested that currents and bathymetry play fundamental roles in this separation, stressing the importance of oceanographic conditions to the population structure of elasmobranchs. Unfortunately, most of the works at national scale are based on samples from the PBC and the GC, while the MTEP is significantly less explored. Few comparative papers with coastal species show an interesting separation between tropical and temperate lineages. Separation occurs at a population level in the scalloped hammerhead shark *S. lewini* (Castillo-Olguín et al., 2012) and at a phylogenetic level in the bedded guitarfish *Zapteryx* spp. (Castillo-Páez et al., 2017). This supports the imperative of developing more extensive research around the MTEP.

#### 2.2.2.4 Finer scale:

Finally, there are a few works at a finer geographic scale (<100 km among several sampling sites), with two patterns arising from these studies. First, the species studies show clear isolation by geographic distance, suggesting that the mobility of costal elasmobranchs is more limited than previously believed. For example, Sandoval-Castillo and Beheregaray (2015) reported high connectivity in the smooth-hound shark from the Northern PBC to the Upper GC (>2000 km). However, they also found low genetic differentiation correlated with geographic distance, and that this correlation is stronger in females than in males. The authors suggest the management of the species should consider a metapopulation with subpopulations and migratory corridors that should be preserved in order to reduce the extirpation or even extinction of the heavily exploited species. Isolation by distance was also found in a more pelagic species, the smooth hammerhead *Sphyrna zygaena*, with adult females and juveniles showing a stronger spatial autocorrelation (Félix-López et al., 2019). These highlight the need for studies that use high-resolution geographic sampling to understand migration patterns relevant for the management of coastal species.

Second, high geographic resolution combined with oceanographic data could allow testing of the adaptive effects of oceanographic factors on the genetic structure of elasmobranchs. More recently, Sandoval-Castillo and Beheregaray (2019) have shown isolation by ecological distance (genetic and environmental distances are directly correlated, independent of geographic distance, Shafer and Wolf, 2013) in one species within the GC. In this region, the shovelnose guitarfish *Pseudobatos productus* has lower genetic connectivity, and, differences in temperature and bathymetry are correlated with the genetic differentiation between three subpopulations. The geographic distribution of each of these three subpopulations overlap with three biogeographic regions previously defined based on the complex oceanography of the GC (Ortega et al., 2010). The authors suggest that these populations are independent management units with demographic differences potentially associated with local adaptations. This should be considered in the management plan for the species, and perhaps, the same pattern could be present in other commercially important species in the region.

### 2.2.3 Demography

The significant population declines and increasing extinction risks over several species of sharks and rays around the world have become a major concern for conservation biologists in the last decade (Dulvy et al., 2014, 2017; Worm et al., 2013). The genetic diversity loss associated with population declines could lead to excessive genetic drift and inbreeding depression. Potentially severe consequences include compromised evolutionary persistence and resilience of species to environmental changes (Frankham, 2015). Genetic studies aiming to elucidate demographic parameters such as connectivity, population size, and genetic diversity are essential for identifying the risks associated with population size declines (Allendorf et al., 2013), and consequently for the management and conservation of elasmobranchs.

Several works in the MPC have used genetic diversity as a rough estimate of population size, however very few have used coalescent or linkage disequilibrium analyses which are better able to infer genetic signals of demographic events. Regardless, a common pattern in the MPC is the effect of the last glacial maximum. For example, Ramírez-Macías et al. (2007) suggested that whale sharks in the GC have passed through a recent population expansion, based only on high haplotype diversity but low nucleotide diversity of sequences in the mitochondrial control region. Coalescent analyses support the idea of a moderately large global population (Castro et al., 2007; Schmidt et al., 2009) with a recent demographic expansion (Vignaud et al., 2014). Sandoval-Castillo and Beheregaray (2015) report a geographic expansion from the GC to the PBC, possibly associated with the function of the GC as refuge during the last glacial maximum. Using a couple of coalescent approaches King et al. (2015) report historical and contemporary small effective population size on the blue shark from the Pacific Ocean. This is an unexpected result for a pelagic shark with high abundance, vagility, and fecundity that comprises a single breeding unit across a wide geographic range in the North Pacific Ocean. However, the authors attribute the low effective population size (Ne), to a population bottleneck during the last glacial maximum and posterior geographic expansion during the actual interglacial period. On the other hand, the scalloped hammerhead in the Eastern Pacific shows evidence of a population bottleneck related with geographic expansion and population subdivisions during the Holocene (Castillo-Olguín et al., 2012; Nance et al., 2011). Meanwhile, the pelagic silky shark has a historically stable population in the NEP, but shows local demographic expansions possibly associated with the onset of warmer currents following the last glacial maximum (Galván-Tirado et al., 2013). Finally, the smooth hammerhead in the North Mexican Pacific has experienced a recent population expansion from a bottleneck which occurred during the Late Pleistocene (Félix-López et al., 2019).

Most of these studies have estimated historical Ne based on coalescent theory and mtDNA sequence. Although the demographic history of a population can determine the relative consequences of modern population declines (van der Valk et al., 2019), contemporary Ne reflects current genetic health and can approximate recent numbers of breeding individuals (Frankham et al., 2010). Contemporary Ne is more relevant for the conservation and management of existing populations. Using genetic data to estimate contemporary abundance relies on the genetic signal associated with inbreeding to calculate Ne, and the equivalence between Ne and abundance (e.g. census size or Nc). This equivalence depends largely on several life history characteristics, which may be difficult to obtain (Ovenden et al., 2016). While no consistent relationships exist across taxa, larger species with long-lived individuals, low fecundity, late maturity and high survivorship of adults more often than not show Ne/Nc ratios approaching one (Dudgeon and Ovenden, 2015; Portnoy et al., 2008) facilitating the demographic interpretation of genetically estimated Ne. These are life history characteristics of most elasmobranchs, making Ne a promising method for assessing trends in population sizes important for the management of sharks and rays (Dudgeon and Ovenden, 2015). As such, an increase in studies estimating and monitoring contemporary Ne in elasmobranch populations should be helpful for improving management decisions in the MPC.

### *2.2.4 Reproductive strategies*

Different reproductive strategies result in different recruitment efficiencies. In the long-term, this can influence population stability and evolutionary resilience to population declines (Lowerre-Barbieri et al., 2017). Thus, knowledge of reproductive strategies is important for implementing management plans, including recovery plans for elasmobranch populations depleted by human activity. Unfortunately, direct observation of the reproductive biology of elasmobranchs is difficult due to their biological characteristics (Compagno et al., 2005; Portnoy and Heist, 2012). Currently, a wide variety of molecular approaches can be used to provide information about the reproductive biology of elasmobranchs, with several examples around the world (Portnoy and Heist, 2012), including a few within the NEP (Larson et al., 2017).

Two main reproductive strategies have been explored in elasmobranchs using genetic data, with the first being female philopatry to mating or nursery areas. The occurrence of this behaviour has been shaped by selective pressures associated with juvenile survival and generational recruitment, and therefore has a significant impact on species resilience (Portnoy and Heist, 2012). The study of genetic female philopatry in elasmobranchs originally emerged because it was the easiest explanation for incongruences in genetic structure between nuclear and mitochondrial markers (Dudgeon et al., 2012). However, the idea was later supported by more complex analyses including comparative analyses between females and males (Sandoval-Castillo and Beheregaray, 2015), and relatedness analyses combined with parental analyses (Feldheim et al., 2014).

In the MPC, several papers studying the population structure of different elasmobranch species have suggested some level of female philopatry, even in species with remarkably different dispersal potentials. Most of these findings have been based solely on the presence of mitochondrial genetic structure (e.g. Castillo-Olguín et al., 2012; Félix-López et al., 2019; Ramírez-Amaro et al., 2017; Sandoval-Castillo and Rocha-Olivares, 2011; Sandoval-Castillo et al., 2004) or the discrepancy between signals of mitochondrial and nuclear differentiation (Castillo-Páez et al., 2017; Daly-Engel et al., 2012). However, these results could also be explained by the intrinsic evolutionary characteristics of mtDNA (Avise et al., 1987). As such, more ad hoc statistical analyses and sampling designs are required to estimate the effect of sex-biased dispersal on the observed population structure (Dudgeon et al., 2012; Portnoy and Heist, 2012). One exception in the MPC is the work by Sandoval-Castillo and Beheregaray (2015), who integrated high geographic resolution sampling and individual-level genetic analysis to determine effective dispersal on the smooth hound shark. The authors used the relatively lower effective dispersal of females compared with males to infer female philopatry. Although these approaches still can confuse between philopatry (actively selecting a site) and limited home range (limited mobility), the authors combine genetic seascape analyses and the presence of mating and nursery aggregation in the GC to support their hypothesis. Perhaps the best proof of philopatry in any shark from the MPC is the work by Jorgensen et al. (2009). In this paper, genetic data was combined with satellite tag data to demonstrate that both sexes preferentially return to a small subset of available coastal sites. However, molecular tools can be powerful enough to be used as permanent identification tools in long-term mark and recapture experiments. If sample size is large enough, kinship and parentage analysis can discriminate as to whether female philopatry occurs in a few individuals or in a whole population (Feldheim et al., 2014; Veríssimo et al., 2017), providing more valuable and reliable information relevant to the management of elasmobranchs.

The second main reproductive strategy that has been analysed with molecular data is polygamy, or more specifically polyandry (multiple mating by females). It is important to determine the prevalence of polyandry, because it can change the effective population size and genetic diversity of a population, with significant demographic consequences (Byrne and Avise, 2012; Karl, 2008). However, due to inherent logistical difficulties, polyandry has been directly observed in few species of elasmobranchs (Pratt and Carrier, 2001; Snelson et al., 2009). Molecular techniques have allowed assessment of the frequency of polyandry by calculating the multiple paternity levels in litters. In fact, polyandry detected using genetic approaches has been reported in over 20 species of elasmobranch (Boomer et al., 2013; Corrigan et al., 2015b; Fitzpatrick et al., 2012; Rossouw et al., 2016) including a few from the NEP (Larson et al., 2017). However, the occurrence and prevalence of multiple paternities in elasmobranchs varies between species, populations, individuals and even reproductive seasons. Polyandry in samples from the MPC has been reported just in the brown smooth-hound shark. Analysing 14 litters and their respective mothers collected from the PBC, Byrne and Avise (2012) report one of the highest occurrence of multiple paternity of any shark (93%). However, depending of the year of collection, only a 0–40% incidence of multiple paternity was observed in 18 litters of the same species collected from California (Chabot and Haggin, 2014). Moreover, Tarula-Marin and Saavedra-Sotelo (2019) determined that genetic monogamy is prevalent in the young grey smooth-hound shark, a very closely related species. It was previously suggested that the differences in polyandry frequency depend on factors such as mother's size, individual behaviour, home rage, philopatric tendencies, population size and post copulatory mechanisms (Boomer et al., 2013; Byrne and Avise, 2012; Chabot and Haggin, 2014; Green et al., 2017; Larson et al., 2011; Lyons et al., 2017). But more comparative analyses (between populations, reproductive seasons and species) are necessary to not just understand these differences in frequency, but also understand why polyandry is present in elasmobranchs.

Intuitively, polyandry has multiple benefits; it can increase fertilization success, counteract inbreeding issues, elevate the genetic diversity of the offspring and therefore relative fitness, and it can increase the effective population size (see Byrne and Avise, 2012; Green et al., 2017). However, the few studies comparing genetically polyandrous against monogamous litters of sharks did not find evidence of higher fertilization success (Portnoy et al., 2007), nor an increase in survival rate or genetic diversity of offspring (Dibattista et al., 2008). Moreover, instead of increasing Ne, polyandry can reduce it when the parental contribution per litter is highly skewed and variable (Karl, 2008). This is the case in most empirical data for elasmobranchs (Chabot and Haggin, 2014; Green et al., 2017; Larson et al., 2011; Rossouw et al., 2016). These apparent contradictions have given support to the convenience polyandry theory, which suggests that females increase receptivity to mating when the relative costs of resistance overcome the cost of mating (Boulton et al., 2018). More empirical testing is needed to rule out other possible evolutionary benefits of polyandry. However, understanding the reasons for polyandry and its demographic effects in elasmobranchs is fundamental to implementing appropriate fisheries management plans, and currently there are few efforts towards the evaluation of mating systems in rays and sharks from the MPC (Saavedra-Sotelo personal communication, 2019).

## 2.3 Taxonomic and geographic bias

We found 47 scientific research articles on elasmobranch genetics using samples from the MPC (Table 1). Below, we describe a few of the most notable patterns regarding the nature of this work. Temporally, a clear exponential growth in genetic research occurs in the last decade, when >69% of the works were produced. This follows the same pattern of increase across

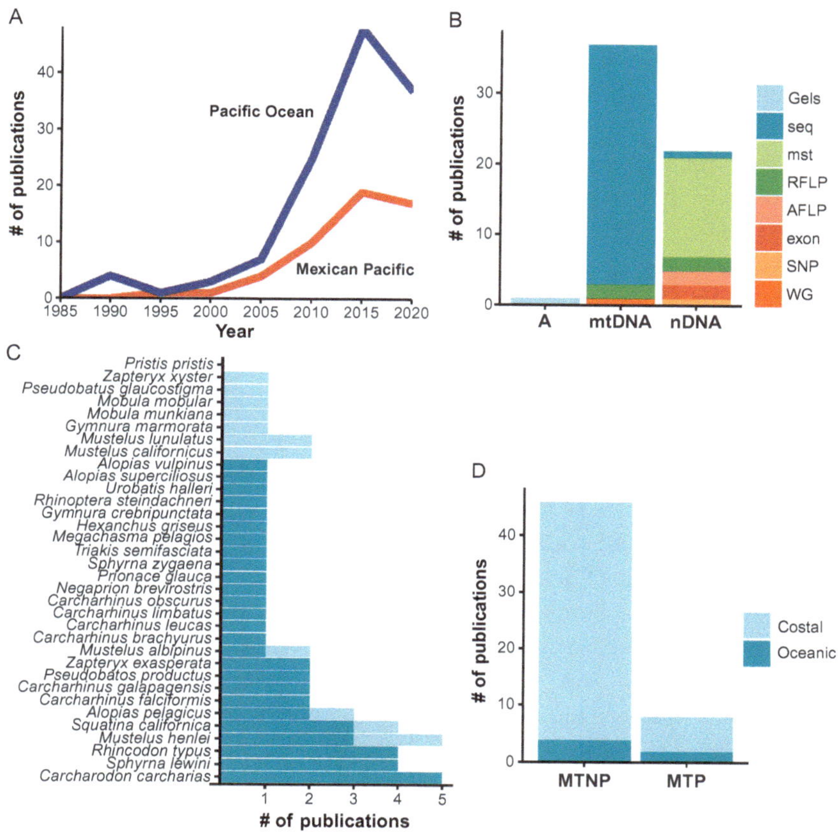


**Fig. 1** Genetic studies in elasmobranchs from the Mexican Pacific Coast (A) Number of publications per year: from 1985 to 2019. (B) Nuber of publications per marker and method: A = allozymes; mtDNA = mitochondrial DNA; nDNA = nuclear DNA; seq = sequences; mst = microsatellites; RFLP = restriction fragment length polymorphisms; SNP = single nucleotide polymorphisms; WG = whole genome. (C) Number of publications per species: light blue = phylogenetic studies with no intraspecific information; dark blue = intraspecific studies. (D) Number of publications per region: MTNP = Mexican Temperate Northern Pacific, MTEP = Mexican Tropical Eastern Pacific.

the entire Pacific Ocean, but at a smaller scale (Fig. 1A). The works are dominated by analysis of mtDNA sequences, followed by microsatellites (Fig. 1B); unfortunately, few papers combine the power of bi-parental and matrilineal markers (26%). The dominance of mtDNA sequences is similar to that observed at a global level for phylogeographic analyses >10 years ago (Beheregaray, 2008). In the MPC, this dominance is still present and reflects the limited availability of economic and technological resources in the region.

More concerning is that half of the genetic research using samples from the MPC includes null or minimal participation of local researchers (Table 1). We strongly believe that investment in the development of new and support of existent local skills (Mexican molecular ecologists and elasmobranch biologists), and international collaborations would catalyse the production of genomic research in the region. In time, this will improve our ability to implement efficient management plans for elasmobranchs at a local, international and perhaps even global scale.

The genetic research in the MPC is also taxonomically biased; in the 47 works reported here, 35 species from 14 families were analysed, representing <30% of the species diversity in the MPC (Ehemann et al., 2018). If we remove the species that were analysed with only a few samples for systematic studies (light grey in Fig. 1C), and from where we cannot obtain intraspecific inferences, this number decreased to 27 species, representing just over 22% of the elasmobranch fauna in the region. The family Carcharhinidae dominates the representation with eight species, followed by sharks from the family Triakidae with seven species (Table 1 Fig. 1C). This is expected, since both families are highly diverse and contain commercially important and abundant species in the MPC, especially from the genera *Carcharhinus* and *Mustelus* (see chapter "Fisheries interactions and the challenges of both targeted and non-targeted take in shark conservation" by Sosa-Nishizaki of volume 84). However, per species, the analyses have a different bias. While most species are the main subject of one study, only 13 species were analysed in two or more studies (Table 1 Fig. 1C). More interestingly, there are three species that were the focus in more than three studies with intraspecific analyses, the white shark *Carcharodon carcharias,* the whale shark *Rhincodon typus* and the scalloped hammerhead shark *Sphyrna lewini* (Fig. 1C). These are arguably three of the most charismatic elasmobranchs in the world. However, no studies exist for the largetooth sawfish *Pristis pristis* or the smalltooth sawfish *Plebejus acmon* (but see Section 3), the most critically endangered species (CITES 2019) in the region (Ehemann et al., 2018). The same is true for the Cortez skate *Beringraja cortezensis*, a species endemic to the GC with a very narrow distribution and uncertain conservation status (Last et al., 2016). These patterns reflect a strong bias for large charismatic and/or commercially important species in the field, as occurs in other types of conservation research (McClenachan et al., 2012). We do not promote a reduction in research for charismatic or economically important species; rather, we draw attention to the fact that conservation genetic research has either completely ignored or only partially analysed around 90% of the elasmobranch fauna in the region. In fact, although all these genetic studies focus on economically important species, no genetic data exist for the other 30 species caught as target or by-catch in the region (CONAPESCA-INP, 2004; see chapter "Fisheries interactions and the challenges of both targeted and non-targeted take in shark conservation" by Sosa-Nishizaki volume 84). We believe that to move from the current overexploitation and high extinction risk to a long-term sustainable management of MPC elasmobranch fauna, genetic research should have a taxonomically broader focus This will require a considerable investment in basic and applied national research, from both local and international entities (e.g. FAO, CONACYT), in addition to the implementation of national and international collaborations. We especially recommend fomenting the use of new sequencing technologies which facilitate the application of traditional and modern genetic approaches, mainly by dropping the cost per sample per locus, decreasing the number of samples necessary for efficient analysis (e.g. Gaughran et al., 2018), and reducing the logistic difficulties implicated with elasmobranch sampling (see Section 3).

Another notable pattern was a geographic bias. First, there are more studies in coastal areas than oceanic regions (Table 1; Fig. 1D). There were just two oceanic sampling sites; the Revillagigedo Archipelago is 390 km southwest from Baja California Peninsula and is considered a hotspot of elasmobranch diversity (Sandoval-Castillo unpublished data), while Guadalupe Island is 260 km offshore of Baja California and is an important aggregation for white sharks (Jorgensen et al., 2009). This bias was expected due to the considerable logistic difficulties implicated in sampling from oceanic areas. But more concerning is the second geographic bias, the difference between the northern and southern MPC. The MPC includes two marine biogeographic realms, the Temperate Northern Pacific (TNP) and the Tropical Eastern Pacific (TEP). Both realms have relatively high elasmobranch diversity (Carrillo-Briceño et al., 2018; Ehemann et al., 2018), supporting important fisheries and tourism activities (see chapters "The economy of shark conservation: The role of ecotourism and citizen science" by Cisneros-Montemayor of volume 84; "Fisheries interactions and the challenges of both targeted and non-targeted take in shark conservation" by Sosa-Nishizaki of volume 84). However, the number of genetic works using samples from the TEP is much smaller than those using samples from the TNP (Table 1; Fig. 1D). The southwest of Mexico has the lowest social and technological development in the country, which restricts the scientific productivity in the Mexican TEP (Díaz-Gómez et al., 2018). But also, the geographic closeness of the Mexican TNP with USA has promoted the elasmobranch research in the region and this is reflected by the number of papers by or in collaboration with American institutions (Table 1). The TEP is an important biogeographic region with high diversity of elasmobranchs (Carrillo-Briceño et al., 2018) and it is evident than the different oceanographic conditions between the TNP and the TEP can separate populations of the same species (Castillo-Olguín et al., 2012) or sister species (Castillo-Páez et al., 2017). Thus, expanding elasmobranch research to the Mexican TEP is imperative not just for delimiting variation among stock, but also for understanding the evolutionary and ecological processes that maintain and produce elasmobranch diversity. With genetic knowledge expanding and sequence technology costs decreasing, conservation genetic/genomic research in the MPC should more easily cover more species and localities, and therefore improve the development of efficient management plans.

## 3. Future Work

In the last decade, high-throughput sequencing technologies (HST) have facilitated genomic research of non-model species. This has made it possible to genotype hundreds of individuals at thousands of loci. It has also allowed the study of variation in gene expression and epigenetic variation without the need for a reference genome. Even whole-genome analysis is now feasible for non-model organisms (Luikart et al., 2019; Primmer, 2009). Large, genome-wide datasets have drastically changed conservation genetics research, reducing its relative cost, expanding the questions it can address, and boosting its power and accuracy (Primmer, 2009). However, these technological advantages have not yet been widely exploited in the MPC, and in fact have not yet been extensively used in elasmobranch research at a global level (but see Corrigan et al., 2017; Pazmiño et al., 2018). Although HST have a lot of potential, we describe below what we believe could be the most relevant applications for conservation biology of sharks and rays in the MPC.

### 3.1 Forensic genomics

Since DNA barcoding can efficiently discriminate between species and even populations of elasmobranchs (Hanner et al., 2016; Rodrigues-Filho et al., 2012; Ward et al., 2008), the approach can be used to help to trace fishery products (including fins and meat) and identify illegal trade of protected species or populations. Conventionally, these approaches use one or two loci sequences (normally the mitochondrial COI and the nuclear ITS2 genes) obtained through PCR amplification from individual specimen genomic DNA (Ward et al., 2008). However, this approach has several challenges. For instance, it requires high concentrations of the DNA template (100–500 ng), and PCR can coamplify pseudogenes (imperfect paralogous copies of a gene) or intracellular endosymbiotic genomes (bacterial, fungal or viral), which can produce spurious variations (Smith et al., 2012). Additionally, intra-individual variability can be present (e.g. heteroplasmy; two or more different organelle genomes in the same individual), and the same genotype can sometimes be detected in sister species due to the presence of foreign DNA (introgressive hybridization) or very recent divergence (incomplete lineage sorting) (Parmentier et al., 2013; Ward et al., 2008). Any of these occurrences can lead to ambiguous or false identification when traditional DNA barcode methods are used. High-throughput sequencing can help to overcome some of these limitations, because millions of sequences from thousands of DNA templates can be generated in parallel. This reduces the required quantity of original DNA template (5–20 ng), allowing analysis of degraded or very small tissue samples (e.g. Sigsgaard et al., 2017). It also facilitates the generation of multiple barcodes for several individuals simultaneously, making the process faster and cheaper (Shokralla et al., 2015). But perhaps most importantly, the analysis of larger numbers of loci can help to identify pseudogenes, endosymbiotic genes, gene introgression, and apparent incomplete lineage sorting (Coissac et al., 2016; Shokralla et al., 2015). This will also increase power to discriminate not just species but also populations or even individuals. This can improve traceability of fishery products including the identification of illegal trades or catch, a key requisite for the better management of elasmobranchs stocks.

## 3.2 Environmental DNA (eDNA)

Most of the research described in this review was done using opportunistic sampling from the by-products of commercial fisheries catches. However, active sampling was used for the two most studied species, the whale shark and the white sharks (Oñate-González et al., 2015; Ramírez-Macías et al., 2007). These protocols involve a spear or a dart with a core modified to take biopsies from free-ranging animals tracked by boat, or sometimes approached by divers. These methods usually require special permits that are not always easy to obtain. For these reasons, active sampling is logistically difficult, expensive and time consuming. Moreover, this sampling can inflict high levels of stress and physical damage to the sampled animals. Fortunately, organisms leave traces of DNA in their environment via faeces, saliva, urine, blood, and skin cells. Environmental DNA (eDNA) is usually degraded and found in low concentrations, making it extremely difficult to recover any signal with traditional genetic methods (Rees et al., 2014). The massive amount of sequence data obtained from HST has made it possible to recover DNA data directly from environmental samples (Shokralla et al., 2012). The use of eDNA removes the need to capture, injure or even observe the target species, making genetic surveys easier, cheaper and faster. In the MPC, this approach can be used in elusive, rare or critically endangered species for simple presence/absence surveys, such as that currently being done in largetooth and smalltooth sawfish (*Pristis pristis* and *P. pectinata*) from the Mexican TEP (Bonfil et al. unpublished data). Recent papers have also shown that eDNA has the potential for population level analyses, such as the estimation of abundance and genetic diversity of shark aggregations (Lafferty et al., 2018; Sigsgaard et al., 2016). This information will help to identify possible aggregation sites, seasonal movements, and perhaps evaluate fisheries interactions; all of which are essential information for better protection, monitoring, and recovery plans of elasmobranch species in the MPC.

## 3.3 Seascape genomics

Seascape genomics is a relatively new research field, which integrates geographic environmental and genomic variation in marine systems to understand the effects of environmental factors on microevolutionary process such as migration, genetic drift and selection (Balkenhol et al., 2019; Riginos et al., 2016). High-throughput sequencing technologies have increased the power of seascape genomics to resolve genetic structure at very fine geographic scales and taxonomic levels. Analysing thousands of markers provides greater resolution of demographic parameters (e.g. migration rates), and facilitates the identification of candidate adaptive loci (Grummer et al., 2019; Riginos et al., 2016). This information can enable identification of cryptic population boundaries associated with breaks in genetic flow due to adaptive divergences, which could be key for the future adaptability of fisheries resources (Grummer et al., 2019; Riginos et al., 2016). Seascape genetic analysis has so far been used to describe cryptic lineages of elasmobranchs from the GC (Sandoval-Castillo and Beheregaray 2019). In the same study, the authors also identify environmental factors driving major adaptive divergences, as well as the oceanographic barriers limiting connectivity between lineages. This information can be integrated with future climate models to predict the effect of environmental change on the evolutionary potential of the species, as has been done in other taxa (e.g. Razgour et al., 2019). We suggest that the power of seascape genomics can be further applied to not just to elucidate species distributions and geographic population structure in the MPC, but also to clarify the relative role of oceanographic factors influencing microevolutionary processes in the region. As such, it will be useful for determining elasmobranchs' evolutionary potential to respond to fisheries pressure and climatic change.

## 3.4 Demographic analysis

The use of HST to produce genome-wide or even whole-genome data will revolutionize the application of genetics to estimate demographic parameters, including effective population size (historical and contemporary) and inbreeding. Inbreeding increases homozygosity, which can cause inbreeding depression (reduced fitness of inbred individuals; Frankham, 2015). Genomic data can directly measure patterns of homozygosity across the genome, providing more precise estimates of individual inbreeding. Moreover, the use of thousands of genetic variations mapped to relative positions in the genome has allowed implementation of new approaches. For example, with whole-genome sequences we can detect runs of homozygosity (ROH), which are long tracts of homozygosity in an individual produced by identical haplotypes inherited from each parent. ROH can be used to estimate inbreeding with very high precision, and also identify inbreeding depression by testing for associations between the presence of ROH and individual fitness traits (Luikart et al., 2019). Even when fitness data are not available, evolutionary constraints can serve as an approximation of genotypic fitness (see van der Valk et al., 2019). This can be applied in elasmobranches to not just estimate inbreeding depression, but also to develop genetic rescue programmes for populations with extremely low evolutionary potential.

As previously mentioned, contemporary Ne is a very important parameter for conservation genetics, and its calculation has always been challenging. However, the use of genomic data has improved the accuracy, precision, and efficiency compared to previous genetic approaches for estimates of contemporary Ne (Larson et al., 2014; Nunziata and Weisrock, 2018). In addition, several methods for demographic history reconstruction, such as approximate Bayesian computation (ABC; Cornuet et al., 2008) or diffusion approximations for demographic inference (∂a∂i; Gutenkunst et al., 2009), show drastically improved statistical power when used with genomic data, even with few individual samples (~5–15). In fact, historical patterns of Ne can be inferred with the whole-genome sequence of a single individual using pairwise sequential Markovian coalescence (PSMC; Li and Durbin, 2011). Genomic data have been used to estimate Ne in two species of sharks from the Pacific Ocean (Pazmiño et al., 2017; Reid-Anderson et al., 2019). However, they have never been used in MPC populations, despite the importance of estimating Ne and its fluctuations for assessing extinction risk. This knowledge will improve conservation management of endangered species and will help to prioritize efforts in a region with very limited resources for conservation.

## 3.5 Gene expression and epigenetics

High-throughput sequencing technologies can be used to sequence all the messenger RNA within a tissue (RNAseq) in order to analyse the expressed fraction of the genome (transcriptome; Wang et al., 2009). This data can be applied de novo (without any genome reference) to any species to capture functional genomic variation, making it a very useful tool for conservation biology. A population comparative transcriptomic approach can help to identify genes responsive to environmental stresses, and at the same time identify genes and gene pathways that are candidates for local adaptation (Brauer et al., 2017; Sandoval-Castillo et al., 2019). This can improve understanding of genetic activity related to habitat fragmentation and environmental change. As gene expression can vary without any molecular change in the genome (phenotypic plasticity), it can also provide a completely different view of the evolutionary effects of small population sizes (Brauer et al., 2017) and acclimation potential (Healy and Schulte, 2018). There are a few studies that used RNAseq to characterize the transcriptome of shark species (Chana-Munoz et al., 2017; Goshima et al., 2016; Machado et al., 2018; Onimaru et al., 2018; Richards et al., 2013). Although these works are important resources for future research, there is only one study that exploited the potential of comparative transcriptomic analyses in elasmobranchs (Lighten et al., 2016; see below).

Some changes in expression can be transgenerational, and therefore even more relevant for conservation. Heritable variations in ecologically important phenotypic traits can occur in the absence of DNA sequence variation via epigenetic modifications of the genome, including DNA methylations and histone modifications (Goldberg et al., 2007). These modifications are expected to appear at a much faster rate than genetic mutation, potentially allowing organisms to adapt to different habitats or respond to climatic change, even in populations with depleted genetic variation and apparently low evolutionary potential. For example, Lighten et al. (2016) showed differential expression between two populations of winter skate *Leucoraja ocellata*. The authors demonstrated that these differences have an epigenetic basis (the differences in expression are correlated with differences in methylation in the genome) and suggested that they have allowed the rapid adaptation of one population to warmer waters. Although it is still unclear as to how far epigenetic adaptations could help to ameliorate population decline, epigenetics is likely to be an important research topic in conservation genomics for elasmobranchs.

## 3.6 Genomic of fisheries

The MPC is a hotspot of elasmobranch biodiversity (Ehemann et al., 2018) but is also one of the most important regions in the world for elasmobranch fisheries (see chapter "Fisheries interactions and the challenges of both targeted and non-targeted take in shark conservation" by Sosa-Nishizaki of volume 84). There are at least 65 species of sharks and rays that are caught, either as target species or by-catch (CONAPESCA-INP, 2004; see chapter "Fisheries interactions and the challenges of both targeted and non-targeted take in shark conservation" by Sosa-Nishizaki of volume 84). However, the genetic data reported here covers just over 50% of these species; moreover, the existing information has not been implemented into any management plan in the region. Considering the global increase in pressure on marine resources, a failure to incorporate genetic information into future management and conservation of elasmobranches from the MPC puts this biodiversity at high risk. Based on the information presented in this review, we suggest focusing genetic efforts in four fundamental areas:

Species identification: Develop forensic genomics protocols that allow identification of species and population of origin of fisheries products (meat and fins). This would reduce inherent uncertainty of landing data and would also help to identify illegal trade of protected species.

Fisheries stock delimitation: Identify the number of genetic stock and management units of elasmobranches in the MPC to estimate appropriate demographic parameters. This will allow the development of species- and even stock-specific sustainable exploitation plans.

Abundance: Calculate modern Ne and migration rates, which are necessary to determine demographic parameters. This will allow us to estimate not just the abundance of a species, but also its harvesting rate and demographic resilience.

Evolutionary response: Develop long-term seascape genomic, comparative expression and genomic diversity studies, which are necessary to understand the evolutionary response of stocks, populations and species to fisheries pressure and climatic change. This information is necessary to estimate evolutionary potential.

These four suggestions will enable the identification of extinction risk, and the implementation of efficient management strategies a regional and global scale.

## 4. Conclusion

The last decade has borne witness to unprecedented growth in conservation genetic research of MPC elasmobranchs, however the field remains understudied and has enormous potential for further development. The genetic tools described here demonstrate the potential to provide information extremely relevant for conservation, and which is unlikely to be obtained by other methods. This should be enough to justify increased investment in genomic research in the region. With HST costs decreasing, genomic applications are now more accessible (see Glenn, 2011, with updates from Glen, 2016 NGS Field Guide: Overview http://www.molecularecologist.com/2016/03/2016-ngs-field-guide-preview/), even to developing areas such as the MPC. In addition, the analysis of thousands of independent markers also reduces the number of samples required to archive high-precision genetic estimates (Gaughran et al., 2018; Li and Durbin, 2011; Willing et al., 2012), and some methods (e.g. eDNA) even avoid the logistic complications intrinsic to sampling elasmobranches, potentially reducing cost even further (Lafferty et al., 2018; Sigsgaard et al., 2016). The implementation of genomic methods will not just increase our power, precision and accuracy for acquiring conservation-related information: it could also change our perspectives regarding the evolutionary potential of small populations, the use of genetic rescue as a conservation tool and the importance of gene expression plasticity in wild populations. This information will be critical for implementing efficient management plans for elasmobranch populations in the face of climatic change, habitat degradation and overexploitation.

However, conservation genomics can be highly technical, making the understanding and communication of basic concepts challenging. Therefore, developing communication strategies that maximize the appropriate uptake of research outcomes by fishery managers and politicians is a fundamental priority. This will enable the translation of genomic data into applicable conservation and management plans. In summary, we encourage investment in genomic research, and the development of conservation geneticists who are not only able to produce genomic results, but also correctly interpret and translate them in plain terms. This will increase our chances of implementing efficient management and conservation strategies that will reduce the extinction risk of the unique elasmobranch biodiversity in the MPC.

## Acknowledgements

We would like to thank Shawn Larson and Chris Chabot for the valuable comments on an early version. We would especially like to thank Luciano Beheregaray, Katie Gates and Nancy Saavedra-Sotelo for their ideas and contributions which improved the manuscript.

## References

Abercrombie, D.L., Clarke, S.C., Shivji, M.S., 2005. Global-scale genetic identification of hammerhead sharks: application to assessment of the international fin trade and law enforcement. Conserv. Genet. 6, 775–788.

Aguilar, L.A., Matthews, S.A., Ayre, D.J., Minchinton, T.E., 2019. Modelling the differences between El Niño and La Niña years and planktonic larval duration on dispersal across the southeast Australian biogeographic barrier. Geo Geogr. Environ. 6, e00074.

Allendorf, F.W., Hohenlohe, P.A., Luikart, G., 2010. Genomics and the future of conservation genetics. Nat. Rev. Genet. 11, 697.

Allendorf, F., Luikart, G., Aitken, S., 2013. Conservation and the Genetics of Populations. Wiley, Blackwell, Oxford.

Applegate, S.P., Soltelo-Macías, F., Espinosa-Arrubarrena, L., 1993. An overview of Mexican shark fisheries, with suggestions for shark conservation in Mexico. In: Conservation Biology of Sharks. 115, NOAA Technical Report NMFS, pp. 31–37.

Attard, C.R., Beheregaray, L.B., Sandoval-Castillo, J., Jenner, K.C.S., Gill, P.C., Jenner, M.-N.M., Morrice, M.G., Möller, L.M., 2018. From conservation genetics to conservation genomics: a genome-wide assessment of blue whales (*Balaenoptera musculus*) in Australian feeding aggregations. R. Soc. Open Sci. 5, 170925.

Avise, J.C., Arnold, J., Ball, R.M., Bermingham, E., Lamb, T., Neigel, J.E., Reeb, C.A., Saunders, N.C., 1987. Intraspecific phylogeography: the mitochondrial DNA bridge between population genetics and systematics. Annu. Rev. Ecol. Syst. 18, 489–522.

Bailleul, D., Mackenzie, A., Sacchi, O., Poisson, F., Bierne, N., Arnaud-Haond, S., 2018. Large-scale genetic panmixia in the blue shark (*Prionace glauca*): A single worldwide population, or a genetic lag-time effect of the "grey zone" of differentiation? Evol. Appl. 11, 614–630.

Balkenhol, N., Dudaniec, R.Y., Krutovsky, K.V., Johnson, J.S., Cairns, D.M., Segelbacher, G., Selkoe, K.A., von der Heyden, S., Wang, I.J., Selmoni, O., Joost, S., 2019. Landscape genomics: understanding relationships between environmental heterogeneity and genomic characteristics of populations. In: Rajora, O.P. (Ed.), Population Genomics: Concepts, Approaches and Applications. Spring, pp. 261–322.

Barker, A.M., Nosal, A.P., Lewallen, E.A., Burton, R.S., 2015. Genetic structure of leopard shark (*Triakis semifasciata*) populations along the Pacific coast of North America. J. Exp. Mar. Biol. Ecol. 472, 151–157.

Beheregaray, L.B., 2008. Twenty years of phylogeography: the state of the field and the challenges for the Southern Hemisphere. Mol. Ecol. 17, 3754–3774. https://doi.org/10.1111/j.1365-294X.2008.03857.x.

Benavides, M.T., Feldheim, K.A., Duffy, C.A., Wintner, S., Braccini, J.M., Boomer, J., Huveneers, C., Rogers, P., Mangel, J.C., Alfaro-Shigueto, J., 2011. Phylogeography of the copper shark (*Carcharhinus brachyurus*) in the southern hemisphere: implications for the conservation of a coastal apex predator. Mar. Freshwater Res. 62, 861–869.

Bernard, A.M., Richards, V.P., Stanhope, M.J., Shivji, M.S., 2018. Transcriptome-derived microsatellites demonstrate strong genetic differentiation in Pacific White Sharks. J. Hered. 109, 771–779.

Bester-van der Merwe, A.E., Gledhill, K.S., 2015. Molecular species identification and population genetics of chondrichthyans in South Africa: current challenges, priorities and progress. Afr. Zool. 50, 205–217.

Bond, M., Valentin-Albanese, J., Babcock, E.A., Heithaus, M., Grubbs, R., Cerrato, R., Peterson, B., Pikitch, E., Chapman, D., 2019. Top predators induce habitat shifts in prey within marine protected areas. Oecologia 190, 375–385.

Boomer, J.J., Harcourt, R.G., Francis, M.P., Walker, T.I., Braccini, J.M., Stow, A.J., 2013. Frequency of multiple paternity in gummy shark, *Mustelus antarcticus*, and rig, *Mustelus lenticulatus*, and the implications of mate encounter rate, postcopulatory influences, and reproductive mode. J. Hered. 104, 371–379. https://doi.org/10.1093/jhered/est010.

Bornatowski, H., Angelini, R., Coll, M., Barreto, R.R., Amorim, A.F., 2018. Ecological role and historical trends of large pelagic predators in a subtropical marine ecosystem of the South Atlantic. Rev. Fish Biol. Fish. 28, 241–259.

Boulton, R.A., Zuk, M., Shuker, D.M., 2018. An inconvenient truth: the unconsidered benefits of convenience polyandry. Trends Ecol. Evol. 33, 904–915. https://doi.org/10.1016/j.tree.2018.10.002.

Brauer, C.J., Unmack, P.J., Beheregaray, L.B., 2017. Comparative ecological transcriptomics and the contribution of gene expression to the evolutionary potential of a threatened fish. Mol. Ecol. 26, 6841–6856. https://doi.org/10.1111/mec.14432.

Byrne, R.J., Avise, J.C., 2012. Genetic mating system of the brown smoothhound shark (*Mustelus henlei*), including a literature review of multiple paternity in other elasmobranch species. Mar. Biol. 159, 749–756.

Caballero, S., Cardenosa, D., Soler, G., Hyde, J., 2012. Application of multiplex PCR approaches for shark molecular identification: feasibility and applications for fisheries management and conservation in the Eastern Tropical Pacific. Mol. Ecol. Resour. 12, 233–237.

Camhi, M.D., Valenti, S., Fordham, S., Fowler, S., Gibson, C., 2009. The Conservation Status of Pelagic Sharks and Rays: Report of the IUCN Shark Specialist Group Pelagic Shark Red List Workshop. IUCN Species Survival Commission Shark Specialist Group, Newbury, UKx + 78p.

Cardeñosa, D., Hyde, J., Caballero, S., 2014. Genetic diversity and population structure of the pelagic thresher shark (Alopias pelagicus) in the Pacific Ocean: evidence for two evolutionarily significant units. PLoS One 9, e110193.

Carrillo-Briceño, J.D., Carrillo, J.D., Aguilera, O.A., Sanchez-Villagra, M.R., 2018. Shark and ray diversity in the Tropical America (Neotropics)—an examination of environmental and historical factors affecting diversity. PeerJ 6, e5313. https://doi.org/10.7717/peerj.5313.

Castillo-Olguín, E., Uribe-Alcocer, M., Díaz-Jaimes, P., 2012. Assessment of the population genetic structure of Sphyrna lewini to identify conservation units in the Mexican Pacific. Cienc. Mar. 38, 635–652.

Castillo-Páez, A., Sosa-Nishizaki, O., Sandoval-Castillo, J., Galván-Magaña, F., Blanco-Parra, M.-D.-P., Rocha-Olivares, A., 2013. Strong population structure and shallow mitochondrial phylogeny in the banded guitarfish, *Zapteryx exasperata* (Jordan y Gilbert, 1880), from the northern Mexican Pacific. J. Hered. 105, 91–100.

Castillo-Páez, A., Sandoval-Castillo, J., Corro-Espinosa, D., Tovar-Ávila, J., Blanco-Parra, M.-D.-P., Saavedra-Sotelo, N.C., Sosa-Nishizaki, O., Galván-Magaña, F., Rocha-Olivares, A., 2017. Cutting through the Gordian knot: unravelling morphological, molecular, and biogeographical patterns in the genus *Zapteryx* (guitarfish) from the Mexican Pacific. ICES J. Mar. Sci. 74, 1630–1638.

Castro, A., Stewart, B., Wilson, S., Hueter, R., Meekan, M., Motta, P., Bowen, B., Karl, S., 2007. Population genetic structure of Earth's largest fish, the whale shark (*Rhincodon typus*). Mol. Ecol. 16, 5183–5192.

Ceballos, G., Ehrlich, P.R., Barnosky, A.D., García, A., Pringle, R.M., Palmer, T.M., 2015. Accelerated modern human–induced species losses: Entering the sixth mass extinction. Sci. Adv. 1, e1400253.

Ceballos, G., Ehrlich, P.R., Dirzo, R., 2017. Biological annihilation via the ongoing sixth mass extinction signaled by vertebrate population losses and declines. Proc. Natl. Acad. Sci. 114, E6089–E6096.

Chabot, C., Allen, L., 2009. Global population structure of the tope (*Galeorhinus galeus*) inferred by mitochondrial control region sequence data. Mol. Ecol. 18, 545–552.

Chabot, C.L., Haggin, B.M., 2014. Frequency of multiple paternity varies between two populations of brown smoothhound shark, *Mustelus henlei*. Mar. Biol. 161, 797–804. https://doi.org/10.1007/s00227-013-2378-2.

Chabot, C.L., Espinoza, M., Mascareñas-Osorio, I., Rocha-Olivares, A., 2015. The effect of biogeographic and phylogeographic barriers on gene flow in the brown smoothhound shark, *Mustelus henlei*, in the northeastern Pacific. Ecol. Evol. 5, 1585–1600.

Chana-Munoz, A., Jendroszek, A., Sønnichsen, M., Kristiansen, R., Jensen, J.K., Andreasen, P.A., Bendixen, C., Panitz, F., 2017. Multi-tissue RNA-seq and transcriptome characterisation of the spiny dogfish shark (*Squalus acanthias*) provides a molecular tool for biological research and reveals new genes involved in osmoregulation. PLoS One 12, e0182756. https://doi.org/10.1371/journal.pone.0182756.

Clarke, C.R., Karl, S.A., Horn, R.L., Bernard, A.M., Lea, J.S., Hazin, F.H., Prodöhl, P.A., Shivji, M.S., 2015. Global mitochondrial DNA phylogeography and population structure of the silky shark, *Carcharhinus falciformis*. Mar. Biol. 162, 945–955.

Coissac, E., Hollingsworth, P.M., Lavergne, S., Taberlet, P., 2016. From barcodes to genomes: extending the concept of DNA barcoding. Mol. Ecol. 25, 1423–1428. https://doi.org/10.1111/mec.13549.

Compagno, L.J., 1990. Alternative life-history styles of cartilaginous fishes in time and space. Environ. Biol. Fishes 28, 33–75.

Compagno, L., Dando, M., Fowler, S., 2005. Sharks of the World. Princeton Field Guides. Princeton University Press Princeton, New Jersey.

CONAPESCA-INP, 2004. Plan de acción nacional para el manejo y conservación de tiburones, rayas y especies afines en México. Comisión Nacional de Acuacultura y Pesca e Instituto Nacional de la Pesca.

Cornuet, J.-M., Santos, F., Beaumont, M.A., Robert, C.P., Marin, J.-M., Balding, D.J., Guillemaud, T., Estoup, A., 2008. Inferring population history with DIY ABC: a user-friendly approach to approximate Bayesian computation. Bioinformatics 24, 2713–2719. https://doi.org/10.1093/bioinformatics/btn514.

Corrigan, S., Huveneers, C., Stow, A., Beheregaray, L.B., 2015a. A multilocus comparative study of dispersal in three codistributed demersal sharks from eastern Australia. Can. J. Fish. Aquat. Sci. 73, 406–415.

Corrigan, S., Kacev, D., Werry, J., 2015b. A case of genetic polyandry in the shortfin mako *Isurus oxyrinchus*. J. Fish Biol. 87, 794–798. https://doi.org/10.1111/jfb.12743.

Corrigan, S., Delser, P.M., Eddy, C., Duffy, C., Yang, L., Li, C., Bazinet, A.L., Mona, S., Naylor, G.J., 2017. Historical introgression drives pervasive mitochondrial admixture between two species of pelagic sharks. Mol. Phylogen. Evol. 110, 122–126.

Dallarés, S., Pérez-del-Olmo, A., Montero, F.E., Carrassón, M., 2017. Composition and seasonal dynamics of the parasite communities of Scyliorhinus canicula (L., 1758) and Galeus melastomus Rafinesque, 1810 (Elasmobranchii) from the NW Mediterranean Sea in relation to host biology and ecological features. Hydrobiologia 799, 275–291.

Daly-Engel, T.S., Seraphin, K.D., Holland, K.N., Coffey, J.P., Nance, H.A., Toonen, R.J., Bowen, B.W., 2012. Global phylogeography with mixed-marker analysis reveals male-mediated dispersal in the endangered scalloped hammerhead shark (*Sphyrna lewini*). PLoS One 7, e29986.

Davidson, L.N., Krawchuk, M.A., Dulvy, N.K., 2016. Why have global shark and ray landings declined: improved management or overfishing? Fish Fish. 17, 438–458.

Dent, F., Clarke, S., 2015. State of the Global Market for Shark Products. FAO Fisheries and Aquaculture Technical Paper: I, *No. 590*, FAO.

Díaz-Gómez, E., Castro-del Ángel, C.A., Santamaría-Hernández, E., Flores-Montiel, D.M., 2018. Indice Nacional de Cienccia. Tecnologia e Innovacion. CAIINNO.

Díaz-Jaimes, P., Reeb, C., Sosa-Nizhisaki, O., O'Sullivan, J.B., 2016. Assessing genetic variation in juvenile white sharks using teeth collected from landfills in Baja California. México. Conserv. Genet. Resour. 8, 431–434. https://doi.org/10.1007/s12686-016-0589-9.

Dibattista, J.D., Feldheim, K.A., Gruber, S.H., Hendry, A.P., 2008. Are indirect genetic benefits associated with polyandry? Testing predictions in a natural population of lemon sharks. Mol. Ecol. 17, 783–795. https://doi.org/10.1111/j.1365-294X.2007.03623.x.

Domingues, R.R., Hilsdorf, A.W.S., Gadig, O.B.F., 2018. The importance of considering genetic diversity in shark and ray conservation policies. Conserv. Genet. 19, 501–525.

Donaldson, M.R., Burnett, N.J., Braun, D.C., Suski, C.D., Hinch, S.G., Cooke, S.J., Kerr, J.T., 2016. Taxonomic bias and international biodiversity conservation research. FACETS 1, 105–113. https://doi.org/10.1139/facets-2016-0011.

Dudgeon, C.L., Ovenden, J.R., 2015. The relationship between abundance and genetic effective population size in elasmobranchs: an example from the globally threatened zebra shark *Stegostoma fasciatum* within its protected range. Conserv. Genet. 16, 1443–1454. https://doi.org/10.1007/s10592-015-0752-y.

Dudgeon, C., Blower, D., Broderick, D., Giles, J., Holmes, B., Kashiwagi, T., Krück, N., Morgan, J., Tillett, B., Ovenden, J., 2012. A review of the application of molecular genetics for fisheries management and conservation of sharks and rays. J. Fish Biol. 80, 1789–1843.

Dulvy, N.K., Fowler, S.L., Musick, J.A., Cavanagh, R.D., Kyne, P.M., Harrison, L.R., Carlson, J.K., Davidson, L.N., Fordham, S.V., Francis, M.P., 2014. Extinction risk and conservation of the world's sharks and rays. eLIFE 3, e00590.

Dulvy, N.K., Simpfendorfer, C.A., Davidson, L.N., Fordham, S.V., Bräutigam, A., Sant, G., Welch, D.J., 2017. Challenges and priorities in shark and ray conservation. Curr. Biol. 27, R565–R572.

Duncan, K., Martin, A., Bowen, B., De Couet, H., 2006. Global phylogeography of the scalloped hammerhead shark (*Sphyrna lewini*). Mol. Ecol. 15, 2239–2251.

Ehemann, N.R., del Valle González-González, L., Chollet-Villalpando, J.G., De La Cruz-Agüero, J., 2018. Updated checklist of the extant Chondrichthyes within the Exclusive Economic Zone of Mexico. ZooKeys 774, 17.

Eitner, B.J., 1995. Systematics of the genus *Alopias* (Lamniformes: Alopiidae) with evidence for the existence of an unrecognized species. Copeia 1995, 562–571.

Feldheim, K.A., Gruber, S.H., DiBattista, J.D., Babcock, E.A., Kessel, S.T., Hendry, A.P., Pikitch, E.K., Ashley, M.V., Chapman, D.D., 2014. Two decades of genetic profiling yields first evidence of natal philopatry and long-term fidelity to parturition sites in sharks. Mol. Ecol. 23, 110–117.

Félix-López, D.G., Bolaño-Martinez, N., Díaz-Jaimes, P., Oñate-González, E.C., Ramírez-Pérez, J.S., García-Rodríguez, E., Corro-Espinosa, D., Osuna-Soto, J.E., Saavedra-Sotelo, N.C., 2019. Possible female philopatry of the smooth hammerhead shark, *Sphyrna zygaena*, revealed by genetic structure patterns. J. Fish Biol. 94, 671–679.

Fitzpatrick, J.L., Kempster, R.M., Daly-Engel, T.S., Collin, S.P., Evans, J.P., 2012. Assessing the potential for post-copulatory sexual selection in elasmobranchs. J. Fish Biol. 80, 1141–1158. https://doi.org/10.1111/j.1095-8649.2012.03256.x.

Frankham, R., 2015. Genetic rescue of small inbred populations: meta-analysis reveals large and consistent benefits of gene flow. Mol. Ecol. 24, 2610–2618.

Frankham, R., Briscoe, D.A., Ballou, J.D., 2010. Introduction to Conservation Genetics. Cambridge University Press.

Gagnaire, P.A., Broquet, T., Aurelle, D., Viard, F., Souissi, A., Bonhomme, F., Arnaud-Haond, S., Bierne, N., 2015. Using neutral, selected, and hitchhiker loci to assess connectivity of marine populations in the genomic era. Evol. Appl. 8, 769–786.

Galván-Tirado, C., Díaz-Jaimes, P., García-de León, F.J., Galván-Magaña, F., Uribe-Alcocer, M., 2013. Historical demography and genetic differentiation inferred from the mitochondrial DNA of the silky shark (*Carcharhinus falciformis*) in the Pacific Ocean. Fish. Res. 147, 36–46.

Gaughran, S.J., Quinzin, M.C., Miller, J.M., Garrick, R.C., Edwards, D.L., Russello, M.A., Poulakakis, N., Ciofi, C., Beheregaray, L.B., Caccone, A., 2018. Theory, practice, and conservation in the age of genomics: the Galápagos giant tortoise as a case study. Evol. Appl. 11, 1084–1093.

Gervais, C.R., Nay, T.J., Renshaw, G., Johansen, J.L., Steffensen, J.F., Rummer, J.L., 2018. Too hot to handle? Using movement to alleviate effects of elevated temperatures in a benthic elasmobranch, Hemiscyllium ocellatum. Mar. Biol. 165, 162.

Glenn, T.C., 2011. Field guide to next-generation DNA sequencers. Mol. Ecol. Resour. 11, 759–769.

Goldberg, A.D., Allis, C.D., Bernstein, E., 2007. Epigenetics: a landscape takes shape. Cell 128, 635–638. https://doi.org/10.1016/j.cell.2007.02.006.

Goodman, M.C., Hannah, S.M., Ruttenberg, B.I., 2019. The relationship between geographic range extent, sea surface temperature and adult traits in coastal temperate fishes. J. Biogeogr. 46, 1438–1450.

Goshima, M., Sekiguchi, R., Matsushita, M., Nonaka, M., 2016. The complement system of elasmobranches revealed by liver transcriptome analysis of a hammerhead shark, *Sphyrna zygaena*. Dev. Comp. Immunol. 61, 13–24. https://doi.org/10.1016/j.dci.2016.03.009.

Green, M.E., Appleyard, S.A., White, W., Tracey, S., Ovenden, J., 2017. Variability in multiple paternity rates for grey reef sharks (*Carcharhinus amblyrhynchos*) and scalloped hammerheads (*Sphyrna lewini*). Sci. Rep. 7, 1528. https://doi.org/10.1038/s41598-017-01416-w.

Grijalva-Chon, J., Kaichi, A., Numachi, K., 2002. Homogeneidad genética en tiburón angelito (*Squatina californica*) del Golfo de California, evidencia por análisis PCR–RFLP de la región control del ADN mitocondrial. Cienc. Mar 17, 37–42.

Grubbs, R.D., Carlson, J.K., Romine, J.G., Curtis, T.H., McElroy, W.D., McCandless, C.T., Cotton, C.F., Musick, J.A., 2016. Critical assessment and ramifications of a purported marine trophic cascade. Sci. Rep. 6, 20970.

Grummer, J., Beheregaray, L.B., Bernatchez, L., Hand, B., Luikart, G., Narum, S., Taylor, E., 2019. Aquatic landscape genomics and environmental effects on genetic variation. Trends Ecol. Evol. 34, 641–654.

Gubili, C., Duffy, C., Cliff, G., Wintner, S., Shivji, M., Chapman, D., Bruce, B., Martin, A., Sims, D., 2012. Application of molecular genetics for conservation of the white shark, Carcharodon carcharias, L. 1758. In: Domeier, M. (Ed.), Global Perspectives on the Biology and Life History of the White Shark. CRC, pp. 357–380.

Gutenkunst, R.N., Hernandez, R.D., Williamson, S.H., Bustamante, C.D., 2009. Inferring the joint demographic history of multiple populations from multidimensional SNP frequency data. PLoS Genet. 5, e1000695. https://doi.org/10.1371/journal.pgen.1000695.

Hanner, R.H., Naaum, A.M., Shivji, M.S., 2016. Conclusion: DNA-based authentication of shark products and implications for conservation and management. In: Naaum, A.M., Hanner, R. (Eds.), Seafood Authenticity and Traceability. Elsevier, pp. 171–179.

Healy, T.M., Schulte, P.M., 2018. Conserved patterns of alternative splicing in response to cold acclimation in fish. bioRxiv 1, 429704.

Hedrick, P.W., Miller, P.S., 1992. Conservation genetics: techniques and fundamentals. Ecol. Appl. 2, 30–46.

Heupel, M., Simpfendorfer, C., 2010. Science or slaughter: need for lethal sampling of sharks. Conserv. Biol. 24, 1212–1218.

Jorgensen, S.J., Reeb, C.A., Chapple, T.K., Anderson, S., Perle, C., Van Sommeran, S.R., Fritz-Cope, C., Brown, A.C., Klimley, A.P., Block, B.A., 2009. Philopatry and migration of Pacific white sharks. Proc. R. Soc. Lond. Ser. B: Biol. Sci. 277, 679–688.

Karl, S.A., 2008. The effect of multiple paternity on the genetically effective size of a population. Mol. Ecol. 17, 3973–3977. https://doi.org/10.1111/j.1365-294X.2008.03902.x.

Keeney, D.B., Heist, E.J., 2006. Worldwide phylogeography of the blacktip shark (*Carcharhinus limbatus*) inferred from mitochondrial DNA reveals isolation of western Atlantic populations coupled with recent Pacific dispersal. Mol. Ecol. 15, 3669–3679. https://doi.org/10.1111/j.1365-294X.2006.03036.x.

King, J., Wetklo, M., Supernault, J., Taguchi, M., Yokawa, K., Sosa-Nishizaki, O., Withler, R., 2015. Genetic analysis of stock structure of blue shark (*Prionace glauca*) in the north Pacific ocean. Fish. Res. 172, 181–189.

Kitamura, T., Takemura, A., Watabe, S., Taniuchi, T., Shimizu, M., 1996. Mitochondrial DNA analysis for the cytochrome b gene and D-loop region from the bull shark *Carcharhinus leucas*. Fish. Sci. 62, 21–27.

Kyne, P.M., Carlson, J.K., Ebert, D.A., Fordham, S.V., Bizzarro, J.J., Graham, R.T., Kulka, D.W., Tewes, E.E., Harrison, L.R., Dulvy, N.K., 2012. The Conservation Status of North American, Central American, and Caribbean Chondrichthyans. IUCN Species Survival Commission Shark Specialist Group Vancouver, BC.

Lafferty, K.D., Benesh, K.C., Mahon, A.R., Jerde, C.L., Lowe, C.G., 2018. Detecting southern California's white sharks with environmental DNA. Front. Mar. Sci. 5, 1–6.

Larson, S., Christiansen, J., Griffing, D., Ashe, J., Lowry, D., Andrews, K., 2011. Relatedness and polyandry of sixgill sharks, *Hexanchus griseus*, in an urban estuary. Conserv. Genet. 12, 679–690. https://doi.org/10.1007/s10592-010-0174-9.

Larson, W.A., Seeb, L.W., Everett, M.V., Waples, R.K., Templin, W.D., Seeb, J.E., 2014. Genotyping by sequencing resolves shallow population structure to inform conservation of Chinook salmon (*Oncorhynchus tshawytscha*). Evol. Appl. 7, 355–369. https://doi.org/10.1111/eva.12128.

Larson, S.E., Daly-Engel, T.S., Phillips, N.M., 2017. Review of current conservation genetic analyses of Northeast Pacific sharks. Adv. Mar. Biol. 77, 79–110. Elsevier.

Last, P.R., White, W.T., Carvalho, M., Seret, B., Stehmann, M., Naylor, G.J., Marshall, L., 2016. Rays of the World. CSIRO.

Lewallen, E.A., Anderson, T.W., Bohonak, A.J., 2007. Genetic structure of leopard shark (*Triakis semifasciata*) populations in California waters. Mar. Biol. 152, 599–609.

Li, H., Durbin, R., 2011. Inference of human population history from individual whole-genome sequences. Nature 475, 493. https://doi.org/10.1038/nature10231.

Lighten, J., Incarnato, D., Ward, B.J., van Oosterhout, C., Bradbury, I., Hanson, M., Bentzen, P., 2016. Adaptive phenotypic response to climate enabled by epigenetics in a K-strategy species, the fish Leucoraja ocellata (Rajidae). R. Soc. Open Sci. 3, 160299. https://doi.org/10.1098/rsos.160299.

Liu, S.Y.V., Joung, S.J., Yu, C.-J., Hsu, H.-H., Tsai, W.-P., Liu, K.M., 2018. Genetic diversity and connectivity of the megamouth shark (*Megachasma pelagios*). PeerJ 6, e4432.

Lowerre-Barbieri, S., DeCelles, G., Pepin, P., Catalán, I.A., Muhling, B., Erisman, B., Cadrin, S.X., Alós, J., Ospina-Alvarez, A., Stachura, M.M., Tringali, M.D., Burnsed, S.W., Paris, C.B., 2017. Reproductive resilience: a paradigm shift in understanding spawner-recruit systems in exploited marine fish. Fish Fish. 18, 285–312. https://doi.org/10.1111/faf.12180.

Luikart, G., Kardos, M., Hand, B.K., Rajora, B.K., Aitken, S.N., Hohenlohe, P.A., 2019. Population genomics: advancing understanding of nature. In: Rajora, O.P. (Ed.), Population Genomics: Concepts, Approaches and Applications. Springer, pp. 3–82.

Lyons, K., Chabot, C.L., Mull, C.G., Paterson Holder, C.N., Lowe, C.G., 2017. Who's my daddy? Considerations for the influence of sexual selection on multiple paternity in elasmobranch mating systems. Ecol. Evol. 7, 5603–5612.

Machado, A.M., Almeida, T., Mucientes, G., Esteves, P.J., Verissimo, A., Castro, L.F.C., 2018. De novo assembly of the kidney and spleen transcriptomes of the cosmopolitan blue shark, *Prionace glauca*. Mar. Genom. 37, 50–53. https://doi.org/10.1016/j.margen.2017.11.009.

Magnussen, J.E., Pikitch, E.K., Clarke, S., Nicholson, C., Hoelzel, A.R., Shivji, M.S., 2007. Genetic tracking of basking shark products in international trade. Anim. Conserv. 10, 199–207.

McClenachan, L., Cooper, A.B., Carpenter, K.E., Dulvy, N.K., 2012. Extinction risk and bottlenecks in the conservation of charismatic marine species. Conserv. Lett. 5, 73–80.

Meijaard, E., Cardillo, M., Meijaard, E.M., Possingham, H.P., 2015. Geographic bias in citation rates of conservation research. Conserv. Biol. 29, 920–925.

Mieras, P.A., Harvey-Clark, C., Bear, M., Hodgin, G., Hodgin, B., 2017. The economy of shark conservation in the Northeast Pacific: the role of ecotourism and citizen science. Adv. Mar. Biol. 78, 121–153. Elsevier.

Nance, H.A., Klimley, P., Galván-Magaña, F., Martínez-Ortíz, J., Marko, P.B., 2011. Demographic processes underlying subtle patterns of population structure in the scalloped hammerhead shark, *Sphyrna lewini*. PLoS One 6, e21459.

Nunziata, S.O., Weisrock, D.W., 2018. Estimation of contemporary effective population size and population declines using RAD sequence data. Heredity 120, 196–207. https://doi.org/10.1038/s41437-017-0037-y.

Oliver, S., Braccini, M., Newman, S.J., Harvey, E.S., 2015. Global patterns in the bycatch of sharks and rays. Mar. Policy 54, 86–97.

Oñate-González, E.C., Rocha-Olivares, A., Saavedra-Sotelo, N.C., Sosa-Nishizaki, O., 2015. Mitochondrial genetic structure and matrilineal origin of white sharks, *Carcharodon carcharias*, in the Northeastern Pacific: implications for their conservation. J. Hered. 106, 347–354.

Onimaru, K., Tatsumi, K., Shibagaki, K., Kuraku, S., 2018. A de novo transcriptome assembly of the zebra bullhead shark, Heterodontus zebra. Sci. Data 5, 180197. https://doi.org/10.1038/sdata.2018.197.

Ortega, E., Álvarez-Borrego, S., Arriaga, S., Renner, M., Bridge, E., 2010. Physical, chemical, and biological oceanography of the Gulf of California. In: Brusca, R. (Ed.), The Gulf of California: Biodiversity and Conservation. University of Arizona Press, Tucson. University of Arizona, pp. 24–48.

Ovenden, J.R., Berry, O., Welch, D.J., Buckworth, R.C., Dichmont, C.M., 2015. Ocean's eleven: a critical evaluation of the role of population, evolutionary and molecular genetics in the management of wild fisheries. Fish Fish. 16, 125–159.

Ovenden, J.R., Leigh, G.M., Blower, D.C., Jones, A.T., Moore, A., Bustamante, C., Buckworth, R.C., Bennett, M.B., Dudgeon, C.L., 2016. Can estimates of genetic effective population size contribute to fisheries stock assessments? J. Fish Biol. 89, 2505–2518. https://doi.org/10.1111/jfb.13129.

Parmentier, I., Duminil, J., Kuzmina, M., Philippe, M., Thomas, D.W., Kenfack, D., Chuyong, G.B., Cruaud, C., Hardy, O.J., 2013. How effective are DNA barcodes in the identification of african rainforest trees? PLoS One 8, e54921. https://doi.org/10.1371/journal.pone.0054921.

Pazmiño, D.A., Maes, G.E., Simpfendorfer, C.A., Salinas-de-León, P., van Herwerden, L., 2017. Genome-wide SNPs reveal low effective population size within confined management units of the highly vagile Galapagos shark (*Carcharhinus galapagensis*). Conserv. Genet. 18, 1151–1163. https://doi.org/10.1007/s10592-017-0967-1.

Pazmiño, D.A., Maes, G.E., Green, M.E., Simpfendorfer, C.A., Hoyos-Padilla, E.M., Duffy, C.J., Meyer, C.G., Kerwath, S.E., Salinas-de-León, P., van Herwerden, L., 2018. Strong trans-Pacific break and local conservation units in the Galapagos shark (*Carcharhinus galapagensis*) revealed by genome-wide cytonuclear markers. Heredity 120, 407.

Pérez Jiménez, J.C., Nishizaki, O.S., Castillo Geniz, J.L., Buth, D., 2005. A New Eastern North Pacific smoothhound shark (Genus *Mustelus*, Family Triakidae) from the Gulf of California. Copeia 2005, 834–845. https://doi.org/10.1643/0045-8511(2005)005[0834:ANENPS]2.0.CO;2.

Pérez-Jiménez, J., Rocha-Olivares, A., Sosa-Nishizaki, O., 2013. Morphological and molecular differentiation of smooth-hound sharks (Genus *Mustelus*, Family Triakidae) from the Gulf of California. J. Appl. Ichthyol. 29, 268–270.

Phillips, N.M., Chaplin, J.A., Peverell, S.C., Morgan, D.L., 2017. Contrasting population structures of three Pristis sawfishes with different patterns of habitat use. Mar. Freshwat. Res. 68, 452–460.

Plank, S., Lowe, C., Feldheim, K., Wilson Jr., R., Brusslan, J., 2010. Population genetic structure of the round stingray *Urobatis halleri* (Elasmobranchii: Rajiformes) in southern California and the Gulf of California. J. Fish Biol. 77, 329–340.

Plata Zepeda, Y.M., Ramírez Romero, P., Sosa-Rodríguez, F.S., 2018. Economic assessment of tourism based on shark-seeing and diving as a more profitable activity than commercial fishing. In: Leal Filho, W., Pociovălişteanu, D., Borges de Brito, P., Borges de Lima, I. (Eds.), Towards a Sustainable Bioeconomy: Principles, Challenges and Perspectives. Springer, pp. 205–230.

Portnoy, D., Heist, E., 2012. Molecular markers: progress and prospects for understanding reproductive ecology in elasmobranchs. J. Fish Biol. 80, 1120–1140.

Portnoy, D.S., Piercy, A.N., Musick, J.A., Burgess, G.H., Graves, J.E., 2007. Genetic polyandry and sexual conflict in the sandbar shark, *Carcharhinus plumbeus*, in the western North Atlantic and Gulf of Mexico. Mol. Ecol. 16, 187–197. https://doi.org/10.1111/j.1365-294X.2006.03138.x.

Portnoy, D.S., McDowell, J.R., McCandless, C.T., Musick, J.A., Graves, J.E., 2008. Effective size closely approximates the census size in the heavily exploited western Atlantic population of the sandbar shark, *Carcharhinus plumbeus*. Conserv. Genet. 10, 1697. https://doi.org/10.1007/s10592-008-9771-2.

Pouca, C.V., Gervais, C., Reed, J., Michard, J., Brown, C., 2019. Quantity discrimination in Port Jackson sharks incubated under elevated temperatures. Behav. Ecol. Sociobiol. 73, 93.

Pratt, H.L., Carrier, J.C., 2001. A review of elasmobranch reproductive behavior with a case study on the nurse shark, Ginglymostoma cirratum. In: Tricas, T.C., Gruber, S.H. (Eds.),

The Behavior and Sensory Biology of Elasmobranch Fishes: An Anthology in Memory of Donald Richard Nelson. Springer Netherlands, Dordrecht, pp. 157–188.

Primmer, C.R., 2009. From conservation genetics to conservation genomics. Ann. N. Y. Acad. Sci. 1162, 357–368.

Ramírez-Amaro, S., Ramírez-Macías, D., Vázquez-Juárez, R., Flores-Ramírez, S., Galván-Magaña, F., Gutiérrez-Rivera, J.N., 2017. Population structure of the Pacific angel shark (*Squatina californica*) along the northwestern coast of Mexico based on the mitochondrial DNA control region. Cienc. Mar. 43, 69–80.

Ramírez-Macías, D., Vázquez-Juárez, R., Galván-Magaña, F., Munguía-Vega, A., 2007. Variations of the mitochondrial control region sequence in whale sharks (*Rhincodon typus*) from the Gulf of California. Mexico. Fish. Res. 84, 87–95.

Razgour, O., Forester, B., Taggart, J.B., Bekaert, M., Juste, J., Ibáñez, C., Puechmaille, S.J., Novella-Fernandez, R., Alberdi, A., Manel, S., 2019. Considering adaptive genetic variation in climate change vulnerability assessment reduces species range loss projections. Proc. Natl. Acad. Sci. 116, 10418–10423.

Rees, H.C., Maddison, B.C., Middleditch, D.J., Patmore, J.R.M., Gough, K.C., 2014. REVIEW: the detection of aquatic animal species using environmental DNA—a review of eDNA as a survey tool in ecology. J. Appl. Ecol. 51, 1450–1459. https://doi.org/10.1111/1365-2664.12306.

Reid-Anderson, S., Bilgmann, K., Stow, A., 2019. Effective population size of the critically endangered east Australian grey nurse shark *Carcharias taurus*. Mar. Ecol. Prog. Ser. 610, 137–148.

Reis, R., Albert, J., Di Dario, F., Mincarone, M., Petry, P., Rocha, L., 2016. Fish biodiversity and conservation in South America. J. Fish Biol. 89, 12–47.

Richards, V.P., Suzuki, H., Stanhope, M.J., Shivji, M.S., 2013. Characterization of the heart transcriptome of the white shark (Carcharodon carcharias). BMC Genomics 14, 697.

Rick, T.C., Erlandson, J.M., Glassow, M.A., Moss, M.L., 2002. Evaluating the economic significance of sharks, skates, and rays (Elasmobranchs) in prehistoric economies. J. Archaeol. Sci. 29, 111–122.

Riginos, C., Crandall, E.D., Liggins, L., Bongaerts, P., Treml, E.A., 2016. Navigating the currents of seascape genomics: how spatial analyses can augment population genomic studies. Curr. Zool. 62, 581–601. https://doi.org/10.1093/cz/zow067.

Rodrigues-Filho, L.F., Pinhal, D., Sodré, D., Vallinoto, M., 2012. Shark DNA forensics: applications and impacts on genetic diversity. In: Caliskan, M. (Ed.), Analysis of Genetic Variation in Animals. InTech, pp. 269–286.

Rodríguez-Correa, H., González-Rodríguez, A., Oyama, K., 2017. Perspectivas de la Ecología molecular en un país megadiverso. Rev. Mex. Biodivers. 88, 3–13.

Rossouw, C., Wintner, S.P., Bester-Van Der Merwe, A.E., 2016. Assessing multiple paternity in three commercially exploited shark species: *Mustelus mustelus, Carcharhinus obscurus* and *Sphyrna lewini*. J. Fish Biol. 89, 1125–1141. https://doi.org/10.1111/jfb.12996.

Sandoval-Castillo, J., Beheregaray, L.B., 2015. Metapopulation structure informs conservation management in a heavily exploited coastal shark (*Mustelus henlei*). Mar. Ecol. Prog. Ser. 533, 191–203.

Sandoval-Castillo, J., Beheregaray, L.B., 2019. Oceanographic heterogeneity drives ecological radiation in elasmobranchs. J. Biogeogr.

Sandoval-Castillo, J., Rocha-Olivares, A., 2011. Deep mitochondrial divergence in Baja California populations of an aquilopelagic elasmobranch: the golden cownose ray. J. Hered. 102, 269–274.

Sandoval-Castillo, J., Rocha-Olivares, A., Villavicencio-Garayzar, C., Balart, E., 2004. Cryptic isolation of Gulf of California shovelnose guitarfish evidenced by mitochondrial DNA. Mar. Biol. 145, 983–988.

Sandoval-Castillo, J., Robinson, N.A., Hart, A.M., Strain, L.W., Beheregaray, L.B., 2018. Seascape genomics reveals adaptive divergence in a connected and commercially important mollusc, the greenlip abalone (*Haliotis laevigata*), along a longitudinal environmental gradient. Mol. Ecol. 27, 1603–1620.

Sandoval-Castillo, J., Gates, K., Brauer, C., Unmack, P., Bernatchez, L., Beheregaray, L.B., 2019. Gene Expression Responses to Heat Stress to Future Climate in Australian Rainbowfishes. Clim. Change. in preparation.

Schmidt, J.V., Schmidt, C.L., Ozer, F., Ernst, R.E., Feldheim, K.A., Ashley, M.V., Levine, M., 2009. Low genetic differentiation across three major ocean populations of the whale shark, *Rhincodon typus*. PloS one 4, e4988.

Schultz, J., Feldheim, K., Gruber, S., Ashley, M., McGovern, T., Bowen, B., 2008. Global phylogeography and seascape genetics of the lemon sharks (genus *Negaprion*). Mol. Ecol. 17, 5336–5348.

Shafer, A.B., Wolf, J.B., 2013. Widespread evidence for incipient ecological speciation: a meta-analysis of isolation-by-ecology. Ecol. Lett. 16, 940–950.

Shokralla, S., Spall, J.L., Gibson, J.F., Hajibabaei, M., 2012. Next-generation sequencing technologies for environmental DNA research. Mol. Ecol. 21, 1794–1805. https://doi.org/10.1111/j.1365-294X.2012.05538.x.

Shokralla, S., Hellberg, R.S., Handy, S.M., King, I., Hajibabaei, M., 2015. A DNA mini-barcoding system for authentication of processed fish products. Sci. Rep. 5, 15894. https://doi.org/10.1038/srep15894.

Sigsgaard, E.E., Nielsen, I.B., Bach, S.S., Lorenzen, E.D., Robinson, D.P., Knudsen, S.W., Pedersen, M.W., Jaidah, M.A., Orlando, L., Willerslev, E., Møller, P.R., Thomsen, P.F., 2016. Population characteristics of a large whale shark aggregation inferred from seawater environmental DNA. Nat. Ecol. Evol. 1, 0004. https://doi.org/10.1038/s41559-016-0004.

Sigsgaard, E.E., Nielsen, I.B., Carl, H., Krag, M.A., Knudsen, S.W., Xing, Y., Holm-Hansen, T.H., Møller, P.R., Thomsen, P.F., 2017. Seawater environmental DNA reflects seasonality of a coastal fish community. Mar. Biol. 164, 128. https://doi.org/10.1007/s00227-017-3147-4.

Smith, W., Bizzarro, J., Richards, V.P., Nielsen, J., Márquez-Flarías, F., Shivji, M.S., 2009. Morphometric convergence and molecular divergence: the taxonomic status and evolutionary history of *Gymnura crebripunctata* and *Gymnura marmorata* in the eastern Pacific Ocean. J. Fish Biol. 75, 761–783.

Smith, M.A., Bertrand, C., Crosby, K., Eveleigh, E.S., Fernandez-Triana, J., Fisher, B.L., Gibbs, J., Hajibabaei, M., Hallwachs, W., Hind, K., Hrcek, J., Huang, D.-W., Janda, M., Janzen, D.H., Li, Y., Miller, S.E., Packer, L., Quicke, D., Ratnasingham, S., Rodriguez, J., Rougerie, R., Shaw, M.R., Sheffield, C., Stahlhut, J.K., Steinke, D., Whitfield, J., Wood, M., Zhou, X., 2012. Wolbachia and DNA barcoding insects: patterns, potential, and problems. PLoS One 7, e36514. https://doi.org/10.1371/journal.pone.0036514.

Snelson, F.F., Roman, B.L., Burgess, G.H., 2009. The reproductive biology of pelagic elasmobranchs. In: Pitcher, T.J., Camhi, M.D., Pikitch, E.K., Babcock, E.A. (Eds.), Sharks of the Open Ocean. Blackwell, pp. 24–53.

Stein, R.W., Mull, C.G., Kuhn, T.S., Aschliman, N.C., Davidson, L.N., Joy, J.B., Smith, G.J., Dulvy, N.K., Mooers, A.O., 2018. Global priorities for conserving the evolutionary history of sharks, rays and chimaeras. Nat. Ecol. Evol. 2, 288.

Steinke, D., Bernard, A.M., Horn, R.L., Hilton, P., Hanner, R., Shivji, M.S., 2017. DNA analysis of traded shark fins and mobulid gill plates reveals a high proportion of species of conservation concern. Sci. Rep. 7, 9505.

Stelbrink, B., von Rintelen, T., Cliff, G., Kriwet, J., 2010. Molecular systematics and global phylogeography of angel sharks (genus *Squatina*). Mol. Phylogen. Evol. 54, 395–404. https://doi.org/10.1016/j.ympev.2009.07.029.

Stevens, J., Bonfil, R., Dulvy, N., Walker, P., 2000. The effects of fishing on sharks, rays, and chimaeras (chondrichthyans), and the implications for marine ecosystems. ICES J. Mar. Sci. 57, 476–494.
Tarula-Marin, O., Saavedra-Sotelo, N.C., 2019. First record of mating system of the gray smooth-hound shark (*Mustelus californicus*) in the Gulf of California: genetic monogamy in young females. J. Mar. Biol. Assoc. U.K. in review.
Trejo, T., 2005. Global phylogeography ofthresher sharks (*Alopias spp.*) inferred from mitochondrial DNA control region sequences M.Sc. Thesis, Marine Science, California State University.
van der Valk, T., Díez-del-Molino, D., Marques-Bonet, T., Guschanski, K., Dalén, L., 2019. Historical genomes reveal the genomic consequences of recent population decline in Eastern gorillas. Curr. Biol. 29, 165–170.e166. https://doi.org/10.1016/j.cub.2018.11.055.
Vella, N., Vella, A., 2017. Population genetics of the deep-sea bluntnose sixgill shark, *Hexanchus griseus*, revealing spatial genetic heterogeneity. Mar. Genom. 36, 25–32. https://doi.org/10.1016/j.margen.2017.05.012.
Veríssimo, A., Sampaio, Í., McDowell, J.R., Alexandrino, P., Mucientes, G., Queiroz, N., da Silva, C., Jones, C.S., Noble, L.R., 2017. World without borders—genetic population structure of a highly migratory marine predator, the blue shark (*Prionace glauca*). Ecol. Evol. 7, 4768–4781.
Vignaud, T.M., Maynard, J.A., Leblois, R., Meekan, M.G., Vázquez-Juárez, R., Ramírez-Macías, D., Pierce, S.J., Rowat, D., Berumen, M.L., Beeravolu, C., 2014. Genetic structure of populations of whale sharks among ocean basins and evidence for their historic rise and recent decline. Mol. Ecol. 23, 2590–2601.
Wang, Z., Gerstein, M., Snyder, M., 2009. RNA-Seq: a revolutionary tool for transcriptomics. Nat. Rev. Genet. 10, 57. https://doi.org/10.1038/nrg2484.
Ward, R.D., Holmes, B.H., White, W.T., Last, P.R., 2008. DNA barcoding Australasian chondrichthyans: results and potential uses in conservation. Mar. Freshwat. Res. 59, 57–71.
Wetherbee, B.M., Cortés, E., 2004. Food consumption and feeding habits. In: Carrier, J.C., Musick, J.A., Heithaus, M.R. (Eds.), Biology of Sharks and Their Relatives. CRC, pp. 225–246.
White, W., Kyne, P., 2010. The status of chondrichthyan conservation in the Indo-Australasian region. J. Fish Biol. 76, 2090–2117.
White, W.T., Corrigan, S., Yang, L., Henderson, A.C., Bazinet, A.L., Swofford, D.L., Naylor, G.J., 2017. Phylogeny of the manta and devilrays (Chondrichthyes: mobulidae), with an updated taxonomic arrangement for the family. Zool. J. Linn. Soc. 182, 50–75.
Willing, E.-M., Dreyer, C., Van Oosterhout, C., 2012. Estimates of genetic differentiation measured by FST do not necessarily require large sample sizes when using many SNP markers. PloS one 7, e42649.
Worm, B., Branch, T.A., 2012. The future of fish. Trends Ecol. Evol. 27, 594–599.
Worm, B., Davis, B., Kettemer, L., Ward-Paige, C.A., Chapman, D., Heithaus, M.R., Kessel, S.T., Gruber, S.H., 2013. Global catches, exploitation rates, and rebuilding options for sharks. Mar. Policy 40, 194–204.
Yates, P.M., Heupel, M.R., Tobin, A.J., Simpfendorfer, C.A., 2015. Spatio-temporal occurrence patterns of young sharks in tropical coastal waters. Estuar. Coasts 38, 2019–2030.
Young, H.S., McCauley, D.J., Galetti, M., Dirzo, R., 2016. Patterns, causes, and consequences of anthropocene defaunation. Annu. Rev. Ecol. Evol. Syst. 47, 333–358.

FSC
www.fsc.org
MIX
Papier | Fördert
gute Waldnutzung
FSC® C083411

Kolibri 360 GmbH
Ferdinand-Jühlke-Straße 7
99095 Erfurt, Deutschland
produktsicherheit@kolibri360.de